A FIELD GUIDE TO THE
WILDLIFE
of the BRITISH ISLES

Alice Tomsett

p

This is a Parragon Book
This edition published in 2006

Parragon
Queen Street House
4 Queen Street
Bath, BA1 1HE, UK

Produced by Atlantic Publishing

Photographs courtesy of
Oxford Scientific Films
(see page 256 for copyright details)

Text © Parragon Books Ltd 2005

A catalogue record for this book is available
from the British Library.

ISBN 1-40544-393-6

Printed in China

CONTENTS

INTRODUCTION

T here is a rich diversity of wildlife in Britain, and the entries in this field guide have been chosen to highlight and celebrate that fact. Nearly two hundred species are included, giving a rich cross-section of British wildlife, from large mammals like deer to all kinds of birds, from butterflies and beetles to amphibians, reptiles and fish.

Identifying an individual species gives a real sense of achievement, and the aim of this book is to encourage readers to look and find out more. The guide makes identification possible by providing information in each entry about where the species can be found, what should be looked or listened for and how the lifestyle of the particular species has been adapted in order for it to continue to exist.

The population size of many of the species included here has varied in the last five decades. Some, such as the black vine weevil, the mink and the muntjac, have prospered and vastly increased in numbers. Other populations, such as those of the large white butterfly and the great crested newt, have severely decreased, normally due to pollution, loss of habitat or an increase in natural predators. Some species, such as the sika deer, have been introduced from abroad; their numbers have steadily increased. Many, like otters and badgers, are now protected by law, which has maintained their numbers. Conservation groups have ensured that some, such as the osprey, have been reintroduced and then protected in order to allow them to thrive.

The guide has been divided into five different sections to allow for easy reference. In addition, sections are grouped thematically. Different types of birds are grouped together, for example, making it easy to look for garden birds, birds of prey or seabirds. All the information has been gathered together to provide a comprehensive guide to each individual species.

Size

Size is an important factor in the identification of wildlife. Among birds, for instance, size will tell you whether the all-black bird you saw only fleetingly was more likely to be a member of the blackbird family or one of the much larger crows. It may be difficult to estimate size over a distance at first, but constant practice will help.

Habitat
The kind of habitat that a species prefers may also be a big clue to identifying it. Habitat is also important because many animals and birds can only thrive in certain conditions, so it is essential that their habitat is protected or the species may die out.

Common and Scientific names
Different species are organized into families that have the same characteristics. Some families have only a few members while others have many. In addition, each individual animal or bird not only has a common name, but also a Latin scientific name. The first part of this is known as the genus, and indicates a closely related group within a family. The second part identifies a particular species. Sometimes an animal or bird will have a third part to its Latin name, which identifies a sub-species, but these differences cannot usually be easily identified in the wild. One advantage of Latin names is that they are the same in any language.

Ethics and Conservation
People must be very aware of their behaviour when out watching wildlife. Always bear in mind that the welfare of the species must be more important than any other consideration and avoid causing any kind of disturbance. Never do anything that might compromise a habitat. Some species have disappeared purely because their habitat has been damaged or destroyed by human intervention. In addition, some of us like to have bird feeders at home; these bring many species right into the garden, but it is important to keep them clean to avoid spreading diseases through the bird population.

It is also polite, and will help others coming along later, if the rights of landowners and other people are considered and observed. Never trespass or cause any damage to private property.

There are many local and national organizations concerned with animals and birds, and many different conservation groups. These all work to preserve important habitats that are under threat and to protect individual species, and amateurs can usually become involved. Whatever you do, remember that watching wildlife should always be enjoyable.

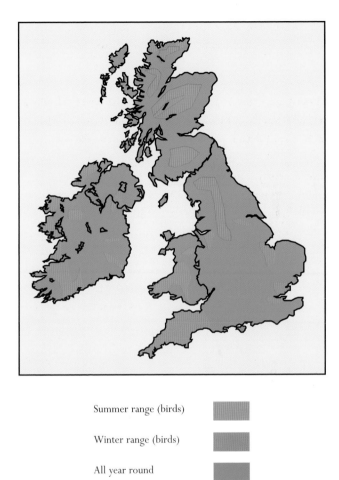

Summer range (birds)

Winter range (birds)

All year round

Range Maps

The same species are not found everywhere, and some are only found in certain areas at specific times. Where relevant, each animal or bird pictured in this book has a range map, which is colour-coded to show when it is most likely to be seen. Birds, in particular, do wander out of their set ranges and storms can blow them off course during migration, but the maps show the most likely candidates within a particular area. All non-migratory species are shown in green.

A Field Guide to the Wildlife of The British Isles

MAMMALS

RABBIT

(ORYCTOLAGUS CUNICULUS)

Originally introduced into Britain from Europe in the twelfth century, the rabbit is now often regarded as a pest. Numbers were drastically reduced in the early 1950s during an outbreak of myxomatosis but have steadily risen again and there are now an estimated 15 million rabbits in Great Britain. Rabbits are very sociable and live together in warrens made up of extensive burrows in the ground. There is a clear hierarchy within the warren, where dominant rabbits mark out their territory and are more successful breeders. Subordinate rabbits do not have territories and mix only with other subordinates. Rabbits are mainly active at night, keeping close to their warren and grazing on vegetation. In winter they may gnaw on bark, which often damages young trees. During the day, while underground, rabbits may eat their soft droppings, which still contain vitamins and proteins. The breeding season extends from January to August; litters are produced at monthly intervals in short burrows away from the buck (male). Three to seven young are born; they are independent within a month and sexually mature at four months.

SIZE Body length 42–60cm.

DESCRIPTION Grey fur, orange at the nape, short white tail with some black on top, long ears and long hind legs. Females are smaller and have narrower heads. Hopping gait. Prominent eyes give good all-round vision. Keen sense of smell.

HABITAT Woodland, dunes, cultivated land.

SOUNDS Scream when captured and thump hind legs when frightened.

SIGNS Pea-sized round droppings.

SIMILAR SPECIES Hares, which have black tips to their ears.

BROWN HARE (LEPUS CAPENSIS)

SIZE Body length up to 70cm. Female has longer head and larger body.

DESCRIPTION Brown fur with long hind legs. Short tail. Orange fur on flank and throat. Its long ears have black tips. In the spring mating season hares often stand on their hind legs, 'boxing' at each other. Large ears and eyes give it an excellent warning system against predators.

HABITAT Grassland, dunes, heaths, woodland.

SOUNDS Usually silent but will scream if injured.

SIGNS Damage to saplings and shoots. Well-worn paths in meadowland.

SIMILAR SPECIES Mountain hare, which is smaller with shorter ears and white or greyish fur.

The hare generally spends the day alone in its form, looking for plants and berries at night. Occasionally it eats its own droppings, which will still contain some essential nutrients. It is capable of running at thirty-five miles an hour if disturbed. The hare breeds at any time and the doe has three or four litters a year. Up to five leverets are produced each time and they are born in the open. Their eyes are open at birth and after feeding from the doe for three weeks they are fully independent. The population of brown hares has declined in number over recent years, which is probably due to the growth in intensive farming.

(LEPUS TIMIDUS) MOUNTAIN HARE

During the day the mountain hare remains in its form and grazes for food at dawn and dusk. In winter it may come down from high areas and shelter behind rocks or in crevices, or in a short burrow dug into the snow. Its main diet is heather but it also eats bilberry shoots and rushes in the summer. In addition, it will also eat its own droppings which still contain some nutrients. Breeding takes place between February and August with up to three litters per year and one or two leverets born each time. Its predators include foxes, eagles and buzzards.

SIZE Body length about 50cm.

DESCRIPTION Blue-grey fur in the summer, which turns to white in the winter, except for the black tips to its ears. Also known as the blue hare.

HABITAT Heather, moorlands and mountains. They are abundant in Scotland and there is a small colony in the Peak District. They are common at all altitudes in Ireland, having replaced the brown hare. These usually stay brown all year and are known as Irish hares.

SIGNS Pathways made through the moors where they have eaten the heather.

SIMILAR SPECIES Brown hare, which is larger with brown fur.

WOOD MOUSE
(APODEMUS SYLVATICUS)

The wood mouse is Britain's most common and widespread mouse, which thrives in a variety of habitats. It often frequents outbuildings but does not leave the same strong smell as the house mouse. A very energetic animal, it is only active at night when it feeds on fruit, seeds and small insects. In the winter months, if food is scarce, it may go into an inactive state similar to hibernation in order to conserve energy. It digs its own burrow system underground where it stores food and rests during the day; it is there that a nest chamber will be built for breeding. From March onwards, the female may produce up to four litters with as many as nine young born each time.

SIZE Up to 11cm. Tail length up to 11cm.

DESCRIPTION Sandy fur with greyish-white undersides. Large ears and eyes. Also known as long-tailed field mouse. Has very large hind feet which help it move rapidly.

HABITAT Gardens, woods, hedgerows, fields.

SIGNS Signs of feeding places in old birds' nests.

SIMILAR SPECIES Yellow-necked mouse, which is larger and more colourful.

HARVEST MOUSE (MICROMYS MINUTUS)

SIZE Body length up to 7cm.
Tail length up to 7cm.
DESCRIPTION Yellowish-red fur
with white undersides. Very
long, scaly, prehensile tail
that is used as an
additional foot. Fur is
darker in winter.
HABITAT Hedgerows and
deciduous woods, usually
restricted to southern
England.
SIMILAR SPECIES Wood
mouse, which is larger.

One of Britain's smallest rodents, the harvest mouse
has the ability to climb very thin stalks. Originally it
lived mainly in cornfields but is now more commonly
found in hedge bottoms, only using cornfields to nest
and feed. It is active during both day and night, which
makes it vulnerable to a wide range of predators. The
harvest mouse feeds mainly on insects, seeds, grain
and soft fruit. Nests at ground level are for winter
use and also for storing food, but the breeding nest is
quite unique. Built well off the ground, it is spherical
and about the size of a cricket ball. This nest is woven
by the female during the later stages of pregnancy
using shredded leaf blades; it is supported by the
grasses and reeds around it. Gestation only takes
seventeen to nineteen days, with three to eight young
produced. Females have up to three litters a year and
build a new nest each time. The young mice are
independent at about sixteen days.

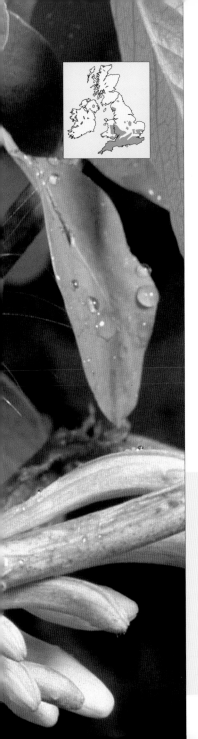

HAZEL DORMOUSE
(MUSCARDINUS AVELLENARIUS)

A dormouse spends much of its time sleeping. It hibernates from October to April, nesting near the ground, and also sleeps in the spring if food is scarce. It is a nocturnal animal and spends the nights looking for hazelnuts and acorns. Pollen from catkins also provides a good source of nutrition. In the summer it uses bark stripped from honeysuckle to build nests above the ground, often in thorny bushes to deter predators. One, occasionally two, litters are produced during the summer months, with up to seven young born each time. They leave the nest when they are a month old.

SIZE Body length up to 8cm. Tail length up to 7cm.
DESCRIPTION Orange-brown fur. Large black eyes, creamy undersides, long fluffy tail. Forelimbs are much shorter than hind limbs.
HABITAT Deciduous woods and hedgerows in southern England and Wales.
SIMILAR SPECIES Edible dormouse, which is larger with grey fur.

HOUSE MOUSE (MUS MUSCULUS)

SIZE Body length up to 9cm. Tail length up to 9cm.

DESCRIPTION Grey-brown fur with long scaly tail. Has greasy fur and emits a strong odour. Has an acute sense of smell.

HABITAT Farms, outbuildings, houses. Widespread throughout the British Isles.

SOUNDS High-pitched squeaks.

SIGNS Holes gnawed in skirting boards, oval droppings.

SIMILAR SPECIES Yellow-necked mouse, which is larger with more sandy-coloured fur.

Originally from Asia, the house mouse thrives wherever there is human habitation. It is a pest, causing damage by gnawing at items such as cables and ruining food supplies with its droppings and urine. Sometimes it carries disease and parasites. It is generally nocturnal and spends the nights searching for food, eating anything but preferring cereals and fats. The house mouse is extremely adaptable and will live in any habitat but usually finds a home in buildings during the winter. House mice breed at a rapid rate, with five to ten litters a year. There are usually five or six in each litter; the young leave the nest at three weeks and the females are sexually mature in six weeks.

GREY SQUIRREL
(SCIURUS CAROLINENSIS)

Originally a native of the United States, the grey squirrel was introduced into Britain in the nineteenth century and has now replaced the native red squirrel throughout most of England and Wales. Found in large numbers, it is a pest that strips bark, damaging trees. It raids birds' nests, will take food left on bird tables and steals from gardens. Its favourite foods include hazelnuts and acorns. It does not hibernate, so a store of food is built up for the winter months. It builds a drey in a tree although nests are also made under roofs in towns. The female has two litters a year, producing up to seven young on each occasion. The young are born blind and naked, are weaned from seven weeks and reach adult size at eight months.

SIZE Body length up to 30cm. Tail length up to 20cm.
DESCRIPTION Grey fur, tinged with orange. Large bushy tail.
HABITAT Woods, gardens, parks.
SOUNDS A churring noise may be heard in trees when it chases away intruders.
SIGNS Remains of nuts may be found on the floor, especially in woodland.
SIMILAR SPECIES Red squirrel, which is smaller with bright chestnut fur.

RED SQUIRREL
(SCIURUS VULGARIS)

The red squirrel is native to Britain but has now been overtaken in numbers by the grey squirrel; competition for food is believed to be the reason for its decline. Mature Scots pine is the preferred habitat of the red squirrel, but it also frequents larch and spruce woods, where it can leap among the high branches and where a good source of cone seeds, pollen, buds and shoots can be found. It also searches amongst deciduous trees for nuts. It does not hibernate but in the autumn will eat well to build up fat reserves; it also stores food for the winter. Dreys are built in trees and are often converted crow or magpie nests. In the mating season, which begins in January, several males will chase one female and a single female is likely to mate with several males. Up to two litters are produced each season and the three or four young are born blind but open their eyes at four weeks. They are reared by the female.

SIZE Body length up to 27cm. Tail length up to 20cm.

DESCRIPTION Red-brown fur, long bushy tail and tufts of fur on ears. It is diurnal with the main peaks of activity at dawn and dusk.

HABITAT Coniferous forests and mixed woodlands.

SOUNDS Makes a 'chuck-chuck' sound if disturbed.

SIGNS Remains of nuts and pine cones fallen onto forest floors.

SIMILAR SPECIES Grey squirrel, which is larger with grey fur.

MOLE
(TALPA EUROPAEA)

The mole is active throughout the year and its system of making tunnels and displacing soil produces the distinctive molehills. A mole only surfaces to collect grass and leaves for its nest, or occasionally to look for food. Each mole has its own system of tunnels which may extend to two hundred metres in length. It regularly patrols the tunnels for food; earthworms are the staple diet and a mole needs to eat half its body weight each day. Once the female has built a nest within the tunnel system, a litter of three or four is produced in April/May, with the young becoming independent after a month. Despite the damage a mole can do, it is also useful, eating insect larvae and aerating the soil with its tunnels.

SIZE Body length up to 15cm. Tail length up to 4cm. Female is slightly smaller.

DESCRIPTION Black, velvety head and body. Pink nose, large forepaws, tiny eyes. Fur is water repellent. Moles have sensitive whiskers and touch sensors on the nose to guide them underground.

HABITAT Gardens, woods, orchards. Lives mostly underground in a system of tunnels.

SOUNDS Usually silent but will squeak when fighting.

SIGNS Molehills and runs can be seen on lawns and pastures.

COMMON SHREW (SOREX ARANEUS)

SIZE Body length up to 8.5cm. Tail length up to 4.7cm.

DESCRIPTION Dark brown back, brown sides and greyish-white undersides. Long snout, very small eyes and short ears. Red tips to its teeth.

HABITAT Marshes, woods, hedgerows, grassland. Shrews can live in very high densities – as many as 70 per hectare.

SOUNDS High-pitched squeaks, often made when its territory is invaded. Twitters as it searches for food.

SIMILAR SPECIES Pygmy shrew, which is smaller.

The common shrew is fiercely territorial and nests independently, except when breeding. It spends most of its time underground, constantly looking for food. Its main diet consists of woodlice, beetles, worms and slugs and it cannot survive for more than three hours without eating. It has to consume nearly the equivalent of its body weight every day. It searches for food in the soil, under leaves or in underground tunnels. The female makes a nest underground or beneath a fallen log and can have up to five litters a year, producing as many as seven young each time. The young shrews are independent at about four weeks. The female often has white hairs on the back of her neck where the male has seized her fur with his teeth during mating. Most shrews do not live longer than a year.

(SOREX MINUTUS) PYGMY SHREW

Britain's smallest mammal, the pygmy shrew weighs between three and six grams, which is less than a ten pence coin. It loses a great deal of energy from body heat, so constantly needs to search for food. It dies of starvation if it fails to eat for more than two hours and consumes the equivalent of its own weight in food every day. It continually grazes over its territory using tunnels dug by other small animals, looking for prey such as beetles, woodlice and spiders. The breeding season lasts from April to August and the young are born in underground nests. Several litters of two to eight are born, and they leave the nest after three weeks.

SIZE Body length up to 64mm. Tail length up to 40mm.

DESCRIPTION Brown fur with pointed, narrow nose and bulbous head.

HABITAT Moors, forest, farmland.

SIMILAR SPECIES Common shrew, which is larger and more aggressive. They rarely mix when living in the same habitat as they seem to alternate their periods of rest and activity.

HEDGEHOG
(ERINACEUS EUROPAEUS)

The hedgehog is a nocturnal animal that hibernates in the winter but will occasionally appear to feed on milder days. It is Britain's only spiny mammal, and a very familiar sight throughout the British Isles. An adult has about five thousand spines on its back and in times of danger it will roll into a ball, relying on the spines to deter predators. It often nests on compost heaps. Its main diet consists of caterpillars, earthworms and beetles but it also eats small mammals and birds' eggs. In winter the hedgehog lives mainly on its fat reserves and will emerge in the spring ready to breed. The young are usually born in early summer and the female produces up to five babies. They are blind initially, with their eyes opening after about fourteen days. At birth they have just a few soft spikes but these increase rapidly until they have about two thousand at six weeks, when they become independent.

SIZE Body length up to 30cm.
DESCRIPTION An insectivore covered with spines with a pointed snout and short legs.
HABITAT Gardens, parks and woodland.
SOUNDS A range of grunts and snorts. Squeals when alarmed.
SIGNS Soft, black droppings with evidence of insect remains.

FOX (VULPES VULPES)

Foxes live in family groups that include the dog (male), the breeding vixen (female) and her cubs. There may also be non-breeding vixens from previous seasons. During the day the groups will rest underground in an earth that has been dug out, often from an old badger sett or rabbit burrow. The fox is a scavenger and will eat anything, including livestock. In May it begins its moult which often results in a scruffy, scrawny appearance during the summer months. Mating takes place between December and February with a litter of about four cubs produced in April. Born blind, they stay in the earth for four weeks and are then taken out at night to play. Eventually the vixen will teach them to hunt, with the young leaving their earth in the autumn or winter to find their own territories.

SIZE Body up to 90 cm. Tail up to 60cm.
DESCRIPTION Reddish fur with a very bushy tail, which normally has a white tip. White undersides. Ears and paws are blackish and eyes are amber.
HABITAT Urban areas and countryside. Thrives in most environments. Mainly hunts for food at night.
SOUNDS High-pitched barks. Vixen screams in breeding season.
SIGNS Distinctive black droppings.

WILDCAT
(FELIS SYLVESTRIS)

The wildcat almost died out in the nineteenth century when it was regarded as vermin, but numbers have gradually increased with the regeneration of Scottish forestry plantations. It hunts alone or in pairs, mainly at dusk and dawn, preying on rabbits, birds, voles and small rodents. A wildcat covers a relatively small territory of about a quarter of a square mile and usually keeps the same mate. The pair will rest during the day and choose a den from which they can survey the surrounding area easily. In the autumn an extra layer of fat is gained to help them survive winter conditions. Mating occurs in March and the litter, usually of four, is born in May. After gradually learning to hunt with their mother the young are weaned at four months.

SIZE Body length up to 90cm. Tail length up to 30cm. Female slightly smaller.

DESCRIPTION Very similar to a domestic tabby cat but with longer fur and larger body. Bushy tail marked with black rings and a blunt end. Body markings are complete stripes. Amber eyes.

HABITAT Remote hills, grouse moors and forests in Scottish Highlands, Scottish Borders and Northumberland.

SOUNDS Wild cry, often heard at night.

SIMILAR SPECIES Domestic tabby cats, which are smaller and have shorter fur.

STOAT (MUSTELA ERMINEA)

SIZE Body length up to 30cm. Tail length up to 80cm.

DESCRIPTION A long, reddish-brown body with white underparts. Long tail with black tip.

HABITAT Woods, farmland, heaths and moors.

SOUNDS Squeaks and chitters.

SIMILAR SPECIES Weasel, which is smaller.

The stoat is a fearless hunter preying on rodents, rabbits and small birds, killing with a rapid bite to the back of the neck. It mainly lives alone in rock crevices or abandoned burrows and will often line its nest with a victim's fur. It moves very quickly, covering up to twenty miles in an hour. In areas that are snow-covered in winter the stoat turns a creamy-white colour (keeping the black tip to its tail) and is then known as an ermine. In slightly warmer climes this colour change may be partial, giving it a patchy appearance. Stoats mate in the summer but the fertilised egg is not actually implanted until the following March, and the young are born in April or May. The litter consists of six or more, initially born blind and helpless, but becoming independent in about six weeks.

(MUSTELA NIVALIS) WEASEL

The smallest of all British carnivores, the weasel is a determined hunter that needs to eat a third of its body weight each day. Its diet mainly consists of mice and voles but it will also prey on rabbits and small birds if the opportunity arises. Each weasel has a territory of about ten to twenty acres and its narrow body enables it to follow and corner animals in their burrows. Its prey is killed by a swift bite to the back of the neck. Females produce litters in April or May and there may be another litter three months later. Four to six young are produced each time and stay with their mother until they are fully grown at twelve weeks. Young weasels are able to breed in their first summer, but most do not live for more than a year; they may be preyed on by larger mammals but are often trapped as vermin or killed on the roads.

SIZE Body length up to 20cm. Tail length up to 50cm. The female is smaller.

DESCRIPTION A long-bodied mammal with white undersides and small brown patches on the throat. Brown tail. Colour remains the same in winter.

HABITAT Found universally wherever there is sufficient and suitable prey.

SOUNDS Squeaks and chitters.

SIGNS Footprints.

SIMILAR SPECIES Stoat, which is larger and changes colour in winter.

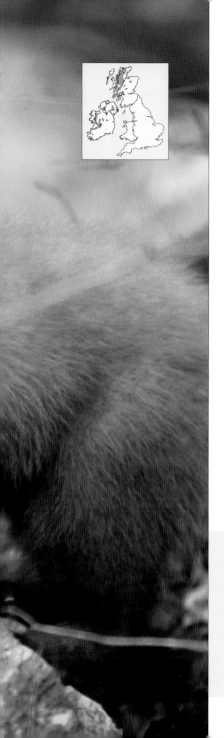

PINE MARTEN
(MARTES MARTES)

The pine marten used to be quite a common sight in Britain but it was widely hunted for its fur and numbers were therefore rapidly reduced. Since 1981 it has been a protected species and the population is gradually on the increase, especially in new plantations. It lives in heavily wooded areas where it can make a den among the roots of a Scots pine or in a hollow tree. It is nocturnal and its diet consists mainly of small birds and mammals, but it will also eat insects and berries. The pine marten is an excellent climber and can withstand falls of up to twenty metres. It mates in July or August but there is a delay in implantation and the female does not become pregnant until the following January. A litter usually consists of three young, born in the den in March or April. After six weeks their eyes open and they then emerge into the open. The young are adult in size by the summer but the family remains together for six months.

SIZE Body length up to 55cm. Tail length up to 27cm.

DESCRIPTION Approximately the size of a cat, it has long reddish-brown fur, a creamy-yellow throat and a very bushy tail. Female is slightly smaller than male. Has a distinctive bounding gait when it runs.

HABITAT Coniferous forest, mixed woodland. Only found in remote areas.

SOUNDS High-pitched chattering in moments of aggression.

SIMILAR SPECIES Weasel, which is much smaller.

POLECAT (MUSTELA PUTORIUS)

Size Body length up to 40cm. Tail length up to 14cm.

Description Dark golden-brown with creamy-yellow underfur. Paws and tail are darker. Dark eye band and whitish chin and ears. Ferrets (domestic polecats) are creamy-white and polecat-ferrets have no mask and a paler forehead.

Habitat Heaths and woodlands in England and Wales.

Sounds Snarls, hisses and growls.

Signs Extremely unpleasant-smelling droppings.

Similar species Domestic ferret, which is darker, and mink, which is smaller.

The polecat used to be known as the foul-mart because of its strong smell. The glands at the base of the tail secrete a foul-smelling scent, which it uses to mark a territory or to defend itself. It was originally hunted for its fur and persecuted due to the threat it posed to livestock; it is hence almost extinct except for a number living in Wales. It will raid a hen house and has the ability to kill all the chickens although it will take only one back to the nest; its name is actually derived from *poule chat* (chicken cat). It is now a protected species and the population is on the increase. There is a lot of interbreeding, which results in many variations of colouring. It hunts mainly at night, preying on small mammals, frogs and birds. Mating takes place between March and May; one litter with five to ten young is produced each year. The young leave the nest when they are two months old.

(MUSTELA VISON) MINK

Introduced into Britain from North America, the mink, with its valuable pelt, was intended to be kept on fur farms. However, it soon escaped and has thrived in the wild, leaving much damage in its wake. Able to hunt on land and in water, the mink indiscriminately kills fish, birds, domestic poultry and small mammals. It is a pest that causes damage worth thousands of pounds to fish farms and crofting communities and has, in the past, devastated colonies of ground-nesting birds. Its presence has also significantly reduced the number of otters and water voles. A nocturnal and solitary animal, the mink has no natural predator. One litter of five or six young is produced each year, and these are born in a den alongside water or in tree roots. They leave the den at about two months.

SIZE Body length 40cm. Tail length 20cm. Female is smaller.

DESCRIPTION Thick, glossy fur that is very dark brown and looks almost black when wet. Small ears and eyes with a pointed snout. Small white spot on lower lip and chin. There are still some animals in existence that were bred to have pale-coloured pelts; these are known as pastel minks.

HABITAT Lakes, rivers, coastlines. Widespread in most of Britain.

SIMILAR SPECIES Otter, which is much larger and paler in colour.

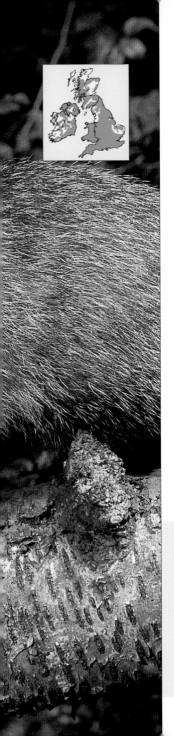

BADGER (MELES MELES)

Now a protected animal under the Badgers Act 1992, the badger is very nocturnal and so is not often seen. It mainly eats worms and small mammals and will eat honey from a bees' nest. It lives in a sett — a large hole dug into the ground and marked by a heap of soil over the top. A sett has lots of interconnecting tunnels with nests and resting places, and the grass and leaves used for bedding are often changed. Latrine pits are also dug nearby. Setts may be hundreds of years old and will be home to about four to twelve adults. They are very sociable and often spend time playing with their young and grooming each other. The badger is also a very territorial animal, keeping to the same paths and tracks throughout its life. It is prone to catching tuberculosis and can live for a long time while infected. In some areas as many as twenty per cent of the badger population may be affected by the disease and many people believe that it is passed on to cattle, although not in any significant number of cases. Badgers usually mate in the spring with implantation delayed until December. Two or three cubs are then born between January and March. They are weaned at twelve weeks and may either stay in the same social group or move to find new territory.

SIZE Body length up to 1 metre. Female is smaller.
DESCRIPTION Dark grey body with distinctive black and white striped head markings, small white-tipped ears. Powerful forepaws with very long claws. Uses rough-barked trees such as oaks to sharpen claws and clean mud from paws. To help recognise badgers from the same group they go through a process called 'musking' by squirting each other with a liquid secreted from a gland under the tail.
HABITAT Deciduous woods.
SOUNDS Noisy animal that grunts, barks and snuffles.
SIGNS Setts, footprints, tufts of hair caught on barbed wire, latrine pits.

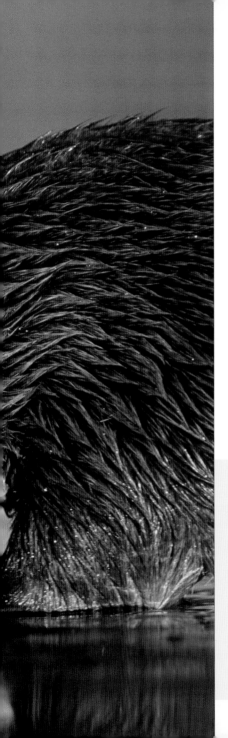

OTTER (LUTRA LUTRA)

Although the otter is now a protected species, numbers have decreased over the years due to pollution, road casualties and loss of habitats. The otter is an extremely strong swimmer and has large lungs that help it to swim and remain underwater for up to four minutes. In addition, it has the ability to slow down its heart rate to conserve oxygen. Its main diet is fish and prey is taken ashore to eat. Otters are very territorial animals and a male will have two or more females within his territory to rear his cubs. The female builds a holt under rocks or tree roots beside the water, and this will often have an underground entrance. A litter of two or three is born and the cubs are taught to swim after about twelve weeks. They are very playful animals and can often be seen frolicking by the water's edge.

SIZE Body length up to 90cm. Tail length up to 45 cm. Female is slightly smaller with lighter head and slimmer neck.

DESCRIPTION Long, streamlined body with powerful, tapered tail. Brown fur and pale undersides. Short legs and webbed feet. Often balances on hind feet and tail to stand up and look round.

HABITAT Coasts, rivers, lakes. They prefer undisturbed water where there is plenty of cover. Very scarce except in Scotland.

SOUNDS Loud chirping noises when chasing each other. Flute-like whistle when hunting.

SIGNS Dark, slimy droppings usually with strong fishy smell. Often left at side of water's edge to mark out territory.

SIMILAR SPECIES Mink, which is smaller with darker fur.

SIKA DEER
(CERVUS NIPPON)

Originally from eastern Asia, the sika deer was introduced into Britain in the second half of the nineteenth century. Many escaped from zoos or parks and all the wild herds are derived from this original source. They live alone or in small groups and graze at dawn and dusk, leaving the sheltered areas where they rest during the day. Rutting occurs from September to November when stags mark their territory by using their antlers to fray tree bark and thrash bushes. Each dominant stag then gathers together a group of hinds. Single calves are born in May or June.

SIZE Height 70–95cm to the shoulder. Weight 30–70kg with the hind smaller than the stag.

DESCRIPTION In the summer has chestnut-brown appearance with spots. Often appears to frown due to lighter hair on the forehead. White tail with dark stripe. Dark edges surround a white rump. Antlers are similar to those of a red deer but are simpler in formation.

HABITAT Mixed woodlands. Most common around Poole in Dorset, the Lake District and the New Forest.

SOUNDS Stag makes a loud, repeated whistle sound in the rutting season.

SIGNS Frayed bark on trees.

SIMILAR SPECIES Red deer, which is larger and does not have spotted markings.

FALLOW DEER
(DAMA DAMA)

The fallow deer has always inhabited the forests of Britain but, more recently, has also been kept in the grounds of stately homes. In the wild it tends to feed at dawn and dusk, spending the day resting. Rutting takes place in October and November when the deer mate; rival bucks will often be seen fighting at this time. The buck's Adam's apple is pronounced and he makes a characteristic 'groaning belch' noise. The male herds together a group of females and marks out his territory with urine. The single fawn is born in June and is usually initially hidden amongst grass or bracken. Twins are occasionally produced.

SIZE Measures 80–95cm to the shoulder and weighs 45–70kg.

DESCRIPTION Several colour varieties. Most common in summer is a chestnut-brown coat with white spots, but some may be pale brown with spots. Long black and white tail and white rump with black border. Males have broad-bladed (palmate) antlers and a prominent Adam's apple. The antlers will vary in appearance regardless of age and are shed each year. In winter the coat is darker with a more uniform appearance.

HABITAT Forests, parks, moorland.

SOUNDS In autumn the buck's 'groaning belch' can be heard when rutting.

SIGNS Droppings and hoofprints.

SIMILAR SPECIES Sika deer, which is slightly smaller.

ROE DEER (CAPREOLUS CAPREOLUS)

SIZE Height 60–75cm to the shoulder, weight 24–30kg. Doe is slightly smaller.

DESCRIPTION Summer coat has a sleek red appearance. It has a distinctive black nose and a white chin. Ears are large with furry inners. Antlers are rough near the base. They are cast in November or December and the new ones that grow in the winter months are protected by a velvet coat – this is rubbed off by May.

HABITAT Woods, forests.

SOUNDS Rasping call during the rut and will bark when frightened.

SIGNS Droppings, tracks.

SIMILAR SPECIES Muntjac deer, which is smaller.

The numbers of roe deer diminished rapidly in the 1500s to the point of extinction in some areas of England, probably because this deer became a major source of food for the peasantry. However, with the boom in forestry and new conservation practices, the population has rapidly increased. The roe deer tends to keep to covered areas and stays alone or in small groups. Grazing on root shoots and shrubs takes place at dawn and dusk. Rutting takes place in July and August after the male has marked his territory. He then mates with any females that enter, chasing off rival bucks. Implantation of the fertilised egg is delayed until December (this is unique to the roe among deer). The kids are born in May or June with twins being common and triplets occurring occasionally. They are hidden among the bracken for the first few days with the doe staying close by and returning regularly to feed them. They come out into the open when they are about two weeks old.

(MUNTIACUS REEVESI) MUNTJAC DEER

Originally from Asia, the muntjac deer was introduced to the Duke of Bedford's Woburn Estate in about 1900. Some were released from there and also from Whipsnade Zoo in 1921. From these two sites in Bedfordshire the population has gradually increased and is continuing to spread through England, southern Scotland and parts of Wales. It is a very hardy animal, active by day and night and it feeds on grasses and brambles as well as plants such as ivy and yew. Like all deer, the muntjac has scent glands with secretions that are believed to be a form of communication. They have forehead glands and mark their territory by rubbing the ground or trees to leave a scent. The buck establishes a territory that includes the home of several does. Unusually he will fight with his teeth, not his antlers. The female can conceive a few days after fawning, has no fixed breeding season and may give birth every seven months – all factors that have, no doubt, aided their survival and their increase in numbers.

SIZE The smallest of all British deer, it measures 43–46cm to the shoulder. Weight 11–16kg

DESCRIPTION Glossy red-brown summer coat and a characteristic rounded rump. Some white markings around the edge of the tail. Short antlers and upper jaw that has fang-like teeth. V-shaped ridge on forehead marked with dark stripes. Antlers are cast in May and grow again in the summer months.

HABITAT Woodland and scrub, sometimes seen in urban areas.

SOUNDS The barking noise it makes also gives it the name of the 'barking deer'.

SIMILAR SPECIES Roe deer, which is larger.

RED DEER
(CERVUS ELAPHUS)

The red deer is Britain's largest land mammal. This native animal is active during the day and night but tends to be more nocturnal in areas where hunting takes place. Stags and hinds live in separate herds except during the rutting season in the autumn. In woodland it lives in fairly small groups but it travels in much larger herds in the highlands, moving uphill to graze during the day and sheltering further down at night. It eats heather, lichen and grasses but, although very hardy, can often die in severe weather despite its ability to scrape through snow in order to find food. There are about quarter of a million red deer in Scotland and these numbers can limit the regeneration of vegetation and trees. Mating takes place in autumn; in the spring, the hind separates to a quiet place where the single calf (occasionally two) is born in May or June. It is initially covered in white spots and can run within eight hours but does not feed itself until it is about nine months old. It stays with its mother until the following autumn. The red deer is sexually mature at three years.

SIZE Male up to 112cm to the shoulder. Weight between 90–190kg. Female 42–54cm, weight 57–115kg.

DESCRIPTION Dark red or brown with buff markings to inner thighs, rump and undersides. In winter is a darker brown or grey with lighter markings in the other areas. Male (stag) grows large, branching antlers. The number of points increases with age until they reach twelve, when the stag is known as a royal.

HABITAT Forests, woods, mountains and moorland. Mainly found in Scotland, the Lake District, Exmoor, the Quantocks and New Forest.

SOUNDS Females bleat, males roar.

SIGNS Droppings, footprints.

SIMILAR SPECIES Sika deer, which is smaller with spots.

GREATER HORSESHOE BAT
(RHINOLOPHUS FERRUMEQUINUM)

Although once common in southern England, the greater horseshoe bat is now in danger of extinction, mainly due to a fall in food sources and the contamination of food by pesticides. It eats insects such as the dung fly, beetles and spiders and only hunts after sunset. It prefers a warmer climate and so can also be affected by weather patterns. Winter hibernation takes place in caves, mines or tunnels; anywhere large and humid. The greater horseshoe bat mates in the autumn or winter with fertilisation taking place in the spring. One offspring is born in July and leaves the roost after five weeks. The female is fertile when she reaches her third year.

SIZE Total length 11cm. Wingspan 34–39cm. Male is smaller.

DESCRIPTION Broad, rounded wings, horseshoe-shaped 'nose leaves' (skin that protrudes from the sides of the nostrils) which are used for echo location. Does not have the pointed growth (tragus) in the inner ear common to most bats. Hangs by gripping surfaces with its claws and then has to bend knees to release grip.

HABITAT Wooded areas, often near water. Only found in south-west England and south Wales.

SOUNDS Emits very high-pitched sounds and uses the echo to find food.

SIMILAR SPECIES Lesser horseshoe bat, which is smaller.

COMMON LONG-EARED BAT (PLECOTUS AURITUS)

SIZE Total length 11cm. Wingspan 23–28cm.

DESCRIPTION Huge oval-shaped ears measuring nearly 3cm that join together at the base. Light brown or yellowish fur. The ears are very sensitive organs that are part of the echo-location system.

HABITAT Woods, house roofs. Will avoid exposed areas. Found all over Britain except in the far north of Scotland.

SOUNDS Although too high-pitched for the human ear to hear, the sound bounces back from objects enabling the bats to source food. The echo indicates the distance and the direction of an object.

SIMILAR SPECIES Grey long-eared bat, which is very slightly larger with deeper grey fur.

The common long-eared bat is nocturnal, being most active between sunset and sunrise, but is occasionally seen during the day. It flies with its ears fully erect, often in very confined spaces such as between tree branches. It can hover in an almost vertical position, enabling it to catch insects and larvae resting on foliage which form most of its diet. It mainly roosts in trees, although it will breed in attics. It normally hibernates from November to March in a cave or mine, often alone. Occasionally these bats will hibernate in summer, but in the roost. Breeding females often cluster together away from the males when nursing. They produce one offspring each year in June or July.

(PIPISTRELLUS PIPISTRELLUS) PIPISTRELLE BAT

The pipistrelle is Britain's smallest and most common bat, living in colonies that sometimes number into the thousands. It emerges from the colony just before sunset and has a very rapid, jerky flight pattern with lots of twists as it hunts. The pipistrelle's diet mainly consists of gnats, moths and caddis flies, and so it is often seen near water. It eats small insects while in flight but larger ones while it roosts. It hibernates from November to March in cool, dry places like old trees or churches. Colonies are mixed when mating takes place, but fertilisation is delayed until April. In summer the female lives in a separate colony with the one or two young that are produced in June. These are born naked, so warmth is essential for survival.

SIZE Body length 5cm. Wingspan 19–25cm.

DESCRIPTION Very tiny with narrow wings and broad, short ears. Colour varies from grey-brown to orange-brown. Flat broad head, blunt muzzle, wide mouth.

HABITAT Woods, farmland, caves, trees, churches, roof spaces. Often found in areas of new housing as they like to roost in summer behind tile-hung walls or weather boarding.

SOUNDS Ultrasonic calls, squeaking.

SIMILAR SPECIES Pygmy pipistrelle, which calls at lower frequency.

A Field Guide to the Wildlife of The British Isles

BIRDS

GREAT CRESTED GREBE (PODICEPS CRISTATUS)

Size 46–50cm

Description Largest of the grebe family, in summer it has a double crest on top of the head, which is black and chestnut. Body is grey-brown with white undersides, a white neck and orange beak. In winter it is greyer in colour and the crest is reduced.

Habitat Ponds, reservoirs and shallow lakes. Widespread except for north-west Scotland. Some spend winter in sheltered coastal regions.

Voice Guttural honks and snorts.

Flight White patches on the wings are clearly visible, the wings beat very quickly and the body appears slightly hump-backed.

Similar species Red-necked grebe, which is smaller and darker with a reddish neck.

Just prior to the breeding season, both the male and female great crested grebe develop elaborate crests, which are fully raised during the courtship display. They shake their heads and rise up out of the water, standing breast to breast and presenting each other with pieces of vegetation. The nest is built by heaping water plants together, building it up from the bottom of the water or anchoring it to branches or stems. About four eggs are usually produced which are whitish and elongated in shape. They take three and a half to four weeks to hatch, producing chicks with distinctive black stripes. These are fed fish and insects by their parents for at least ten weeks. The young are carried on the back of either parent.

(SULA BASSANA) GANNET

Gannets nest in extremely large colonies (known as gannetries) of up to 60,000 pairs. They choose wide ledges or sloping rocks and the nests are evenly spread out. The gannet is, however, extremely aggressive and attacks any bird straying into its territory. It feeds on fish and drops into a vertical flight from as high as thirty metres to catch its prey. The nest is made from a large heap of seaweed and materials such as feathers. Usually a single, elongated egg with a chalky coating is produced. Both parents take it in turns to cover the egg with their webbed feet, carrying out incubation. The chick emerges after about forty-four days and the immature gannet is dark with white spots.

SIZE 86–94 cm. Wingspan 180–190cm.

DESCRIPTION Large white bird with buff-yellow head and neck. Black wing tips and legs. Long bill which turns down slightly at the tip.

HABITAT Widespread around British coastline, especially in the north-west.

VOICE Loud barking sounds, 'urrahs' and 'aarrhs'. Silent at sea.

FLIGHT Shows pointed tail and long, narrow wings exhibiting the black tips. Slow wing beat with frequent glides.

CORMORANT
(PHALACROCORAX CARBO)

Cormorants are members of the pelican family and breed in close colonies, often consisting of thousands of birds. Large nests, made from twigs and seaweed, are sited on cliff tops, ledges and even in trees inland. It eats mainly eels and flatfish, which are caught by diving and then swimming underwater. The most common stance of the cormorant is upright on a rock with its wings spread out to dry; their feathers are less water-repellent than those of other water birds. Three or four eggs are usually laid which are incubated for a month. Both parents feed the young with regurgitated food.

SIZE 80–100cm.

DESCRIPTION Brown-green plumage above, black undersides. Neck and head are black with white patches on cheeks and chin. Long yellow bill that turns down at tip. White patch on thigh during breeding season. Toes are all webbed.

HABITAT Coastal areas and inland waters.

VOICE Deep guttural grunts can be heard on breeding ground.

FLIGHT Head held high – similar to that of a goose.

SIMILAR SPECIES Shag, which is smaller.

TEAL (ANAS CRECCA)

SIZE 34–38cm.

DESCRIPTION Male has green eye patches with white surrounds, a chestnut head and yellow belly. Wings are grey with a green patch. Female is a more dappled brown with a green and black wing patch.

HABITAT Present all over the British Isles with the largest populations in the winter season. Prefers reed-fringed pools and lakes; also estuaries in the winter.

VOICE Female quacks harshly while male has a low 'krit' and a bell-type call.

FLIGHT Fast and erratic; they often fly in crowded clusters with no set pattern.

SIMILAR SPECIES Wigeon, which is larger with cream patches on the male's head.

The teal is Britain's smallest duck. It is a very wary bird that is often difficult to see. It feeds on the surface when swimming or in the shallows, which is a common characteristic of a 'dabbling' duck. Its main diet consists of crustaceans, grain, water plants and insects. The teal population is often low but is increased by migrant birds in the winter. During courtship, the male shows the colourful feathers on his head and tail. The nest is always well hidden amongst vegetation such as reeds, and eight to twelve eggs are laid in May.

(ANAS PLATYRHYNCHOS) MALLARD

The mallard is Britain's most common surface-feeding duck. It lives in a variety of habitats and many have become 'tame' when living near human populations. It has the ability to fly almost vertically into the air from a position on the water surface. Its broad beak allows it to filter food from the water, and it eats almost anything from insects to berries and plant matter. Its feet are webbed and when it walks it has a very distinctive waddle. Nests are made from vegetation and are sited near water but are well hidden. The drake gives a bold display during the courtship period and several will pursue one female; she eventually chooses the one that attracts her most. From eight to twelve pale green eggs are laid between March and May and are incubated for four weeks.

Size 58–60cm

Description Female is a dull buff and brown with a greenish-yellow beak and a violet-blue wing patch. Male has a bottle-green head, a white collar and maroon breast, yellow bill, greyish body and black curly tail feathers.

Habitat Rivers, lakes, ponds and estuaries. Widespread throughout the British Isles.

Voice Female makes classic 'quacking' sound. Male has a more subdued 'raarb' voice.

Flight Blue-purple wing stripe shows when flying.

Similar species Shoveler, which is slightly smaller with browner plumage.

SHOVELER (ANAS CLYPEATA)

SIZE Up to 51cm.

DESCRIPTION Male has bottle-green head, white chest and chestnut belly. Has a very large spoon-type bill, common to both sexes. Female has a speckled brown head and body.

HABITAT Marshes, ponds and lakes with muddy shallows.

VOICE Male has double 'tuk-tuk' quack while female has quiet quack.

FLIGHT Rapid wing beats with tail down and head up.

SIMILAR SPECIES Mallard, which is slightly larger.

The long, spoon-shaped bill of the shoveler has hundreds of little comb-like 'teeth' to filter food from the water. It sifts a large amount of water to find seeds, buds, algae and small molluscs and will also eat crustaceans and tadpoles. It nests in a hollow in the ground, lined with grass and down; eight to twelve greenish-yellow eggs are laid in April or May. They are incubated for just under four weeks by the female and the ducklings are able to fly by the time they are seven weeks old. Resident birds are joined by winter migrants.

(AYTHYA FULIGULA) TUFTED DUCK

Now our most common diving duck, the tufted duck happily inhabits city parks and will tamely accept bread offered to it from passers-by. Its favourite food is the zebra mussel but it will also eat insects, amphibians and fish. When feeding, it dives into the water and then swims using its feet, keeping the wings closed and remaining underwater for up to twenty seconds. It nests on the ground near water; the nest is carefully hidden in vegetation. Usually eight to ten olive-coloured eggs are laid in May or June.

SIZE 40–47cm.

DESCRIPTION Male has dark upper plumage and pure white flanks, with a purple head. Female is brownish all over. Both have a drooping tuft at the back of the head and golden eyes.

HABITAT Any freshwater habitat – widespread throughout Britain.

VOICE Variety of quiet calls; a grating 'currah' sound.

FLIGHT Wing beats are rapid and it shows a white bar running along the length of the wing.

SIMILAR SPECIES Scaup, which has more white at the base of the bill.

MUTE SWAN
(CYGNUS OLOR)

The mute swan uses its long neck to search for underwater plants, which make up most of its diet and are supplemented by worms and insects. Any gravel sucked in with the food contributes roughage. Mute swans pair in the autumn when they are between two and four years old and stay with the same mate for life unless they fail to breed. Once the breeding ground has been chosen a large nest, measuring up to four metres in diameter, is made of a mound of water plants and sited in an isolated spot. The male is extremely territorial during the breeding season and can become very aggressive. Eggs are laid on alternate days for up to twelve days any time between April and July; the female mainly incubates them. They hatch after thirty-six days and, after a further five, the young are able to leave the nest during the day. Immature birds, in their brown plumage, will walk in single file behind a parent and occasionally ride on the female's back when she is on the water. They fly at four months and usually leave the nest the following spring. Non-breeding birds live together until they are old enough to find mates.

SIZE 150cm on average. Wingspan up to 2m.
DESCRIPTION White plumage, long curved neck and orange beak with black knob at the base. Black webbed feet.
HABITAT Widespread except in far north-west of Scotland. Prefer large areas of still water.
VOICE Will hiss or grunt if angry, otherwise no hoot – hence the name.
FLIGHT Powerful with outstretched neck. Flight feathers produce hum when beating.
SIMILAR SPECIES Bewick swan, which is smaller with straight, shorter neck and is much less tame than the mute swan preferring to live in the 'wild'. Mainly found around the Wash and the Severn Estuary.

CANADA GOOSE
(BRANTA CANADENSIS)

The Canada goose was originally introduced to Britain from America in the seventeenth century to populate formal lakes in parkland. Since then both the population and the range of habitats have continued to increase. Attempts to breed the Canada goose as a target for shoots failed, as it is a very tame bird with a low flight and irregular flying times. Its nest is built from plant material, usually by the water's edge. Up to five eggs are laid and the young goslings are able to fly after nine weeks. They stay with their parents until the following spring.

SIZE 90–100cm.

DESCRIPTION Large bird with black head and neck and a white patch on cheeks and chin. Dark brown upper plumage with paler undersides and a white patch below the tail.

HABITAT Large expanses of fresh water.

VOICE Very hoarse, trumpet-like 'aa-honk' sound.

FLIGHT Long neck and very deep wing beats. Black rump and white band on the tail are visible in flight.

SIMILAR SPECIES Brent goose, which is much smaller with darker plumage.

SPARROWHAWK (ACCIPITER NISUS)

SIZE 28–38cm. Wingspan 60–75cm.
DESCRIPTION Dark grey upper parts
with grey crown. Closely barred, red
underparts. Female is larger with
duller brown plumage. Yellow legs.
HABITAT Farmland, woodlands.
VOICE Rapid 'keck-keck-keck'.
FLIGHT Short, rounded wings and a
long tail. Often swoops very low
when looking for prey.
SIMILAR SPECIES Goshawk, which is
larger.

The sparrowhawk is a bird of prey that
ambushes mainly small birds by suddenly
swooping down on them. It can often outfly
its victims and is able to catch them in mid-
flight as well as when they are stationary. It
seizes prey in its long black talons. Courtship
begins in February as the sparrowhawks
circle high above the trees, and nest building
begins in the second half of March. Nests are
made of sticks and usually sited in trees. In
late April about five eggs are laid, which are
blue-white with red-brown blotches. The
female incubates them while the male
searches for food. The parents separate at the
end of the breeding season and find new
mates the following year.

(BUTEO BUTEO) BUZZARD

A very common bird, there are an estimated 20-30,000 buzzards in Britain at present. Always an inhabitant of open areas, the buzzard is becoming increasingly common in more urban environments where it feeds on road kill. Otherwise it eats mice, rats, birds and young rabbits, swooping down on its victims and seizing them in its talons. Its nest is built of sticks and heather stalks, then lined with softer materials such as moss and grass. Two or three eggs are laid at three- to four-day intervals and hatch after a month, so the young vary in age.

SIZE Up to 55cm. Wingspan 115–125cm.

DESCRIPTION Range from pale grey to brown with paler undersides. Short tail, barred neck. Feet and legs are yellow.

HABITAT Open country, moorland, mountains, urban areas.

VOICE Cat-like mewing.

FLIGHT Tends to soar in the air when the broad-fingered wings can clearly be seen with the short, fanned tail.

SIMILAR SPECIES Honey buzzard, which has narrower head and longer tail.

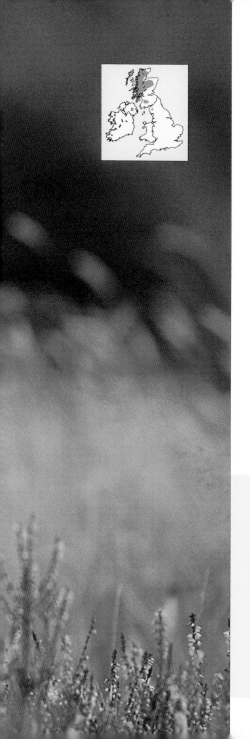

GOLDEN EAGLE
(AQUILA CHRYSAETOS)

A very rare sight, as there are only between two and three hundred breeding pairs of golden eagles in Britain. Each pair needs three thousand acres of hunting ground to survive. The golden eagle preys on hares, ptarmigan and red grouse but is equally capable of killing and lifting a fox from the ground. The nest is made from sticks, heather and grass and is sited on a mountain ledge, sea cliff or in a large tree. Two eggs are laid but usually only one eaglet survives. The immature eagle has white patches on its wings and tail.

SIZE Up to 90 cm. Wingspan 190–250cm.

DESCRIPTION Dark brown plumage with yellow feet and a golden-brown head. Hooked bill.

HABITAT Hebrides, Scottish Highlands, Lake District.

VOICE A shrill yelp or bark; often silent.

SIGNS Glides with limited number of flaps. Long, broad wings with wing tip feathers spread.

SIMILAR SPECIES Sea eagle, which is larger with a white tail.

OSPREY (PANDION HALIAETUS)

SIZE Up to 58 cm. Wingspan 145–160cm.

DESCRIPTION Dark eye streak with white head and a slight crest. Body is dark brown above with white underparts.

HABITAT Present in Scotland and the Lake District in the summer. Likes forest areas near to rivers and large lakes.

VOICE High, repeated whistle; almost chick-like in sound.

FLIGHT Wings are bent making the shape of an 'M'. Flight looks similar to that of a gull, alternately gliding and flapping.

The osprey, which winters in Africa, had disappeared from Britain for fifty years until 1959, when one visiting pair finally reared their young. Under constant protection, this rare bird has gradually increased in numbers in the UK. Its main diet consists of fish and it hunts by hovering high over the water, then plummeting down and entering feet first, often submerging completely. It grasps the fish in its talons, shakes off the water and returns to the nest with its prey. The nest is built from sticks and is used year after year. Constructed normally on top of a pine tree, it is added to each season. Three eggs, generally incubated by the hen, are laid in April or May and hatch after eight or nine weeks.

(FALCO TINNUNCULUS) KESTREL

The kestrel is now Britain's most widespread bird of prey, possibly due to its ability to adapt to urban areas and vary its diet depending on where it lives. In the countryside it shows a preference for rodents but in towns it will prey on small birds. A kestrel hovers while it waits to select its victim and gradually drops in height before swooping on the chosen prey at great speed, grasping it in its talons. It tends to use old nests, often those formerly belonging to crows. Up to five eggs are laid which are incubated by the female; the male will then hunt and bring back food for her.

SIZE 33–35cm. Wingspan 65–80cm.

DESCRIPTION Chestnut back with dark brown spots and grey head. Tail is grey with black band, flecked grey undersides. The upper wings are chestnut with dark tips.

HABITAT Widespread throughout Britain including open ground, farmland, urban areas and motorways.

VOICE Repetitive 'ki-ki-ki'.

FLIGHT Long tail and pointed wings. Has a very distinct hover where it appears rigid except for the quivering of the wing tips.

SIMILAR SPECIES Sparrowhawk, which has a grey back.

PEREGRINE
(FALCO PEREGRINUS)

The peregrine is a merciless hunter that includes a wide range of birds in its prey. It circles in the air while waiting for a suitable victim and then plummets in a stoop that can reach up to one hundred and eighty miles an hour. After killing the prey with a blow from its powerful talons, it will then circle to retrieve the body from the ground where it has fallen. The peregrine nests on the side of a cliff edge and uses virtually no nesting material. Two to four red-brown eggs, blotched with a darker brown, are laid in April or May and take a month to hatch. The young fly in five to six weeks.

SIZE Up to 48cm. Wingspan 95–115cm.

DESCRIPTION Slate grey above, barred with dark grey. White underparts are barred with grey. Top of head and cheeks are nearly black with a moustache-type cheek pattern. Very large eyes give excellent vision.

HABITAT Mountains, moorlands, coastal cliffs.

VOICE 'Kek, kek, kek' sound. Often silent.

FLIGHT Shallow flaps, alternating with glides. Broad wings and relatively short tail shows an anchor shape.

SIMILAR SPECIES Merlin, which is smaller.

RED GROUSE (LAGOPUS LAGOPUS SCOTICUS)

SIZE 37–42cm

DESCRIPTION Rich red-chestnut plumage with white legs and red eye wattles. Female is slightly paler and smaller.

HABITAT Scotland and high moorlands of England and Wales.

VOICE Male makes 'kok-kok-kok' sound and female a 'go back, go back'.

FLIGHT Able to spring up vertically from the heather before gliding away in a low flight. Flaps wings just before landing.

SIMILAR SPECIES Ptarmigan, which has a pure white plumage in winter.

The red grouse is the game bird targeted by hunters on August's 'Glorious Twelfth'. When startled, the bird flies vertically from the heather, making a noisy escape. A red grouse needs young heather plants to eat and more mature plants for cover. It also eats other shoots and seeds, caterpillars and leaves. Regrowth of heather by burning and careful breeding ensures that the numbers of grouse are managed. It nests in a well-concealed, shallow hole and approximately eleven eggs are laid in April or May. These are a creamy colour, blotched with dark brown. Hatching takes place after three and a half weeks and the chicks are able to fly after a further two weeks.

(PERDIX PERDIX) PARTRIDGE

The partridge – also known as the grey partridge – gathers together in family groups (coveys) in September, which is when the main shooting season takes place. When alarmed, the birds rise noisily into the air, often still in groups, and so are very easy targets. The partridge eats insects, caterpillars, green shoots and grass. It pairs in February and the courtship ritual involves the two birds chasing each other. They nest in a hollow in the ground, lined with grass and leaves, concealed by vegetation. The female lays up to twenty olive-coloured eggs. If the male has taken more than one partner, other eggs will be laid in the same nest, making a clutch of forty or more.

SIZE 29–31cm.
DESCRIPTION Head a sandy colour with grey neck and breast. Distinctive horseshoe-shaped brown mark on chest. Reddish-brown tail.
HABITAT Mainly on cultivated land. Also grassland and heaths.
VOICE 'Chirrick, chirrick'.
FLIGHT Whirring and gliding flight which is fast, low and direct.
SIMILAR SPECIES Red-legged partridge, which has white cheeks and a red bill.

PHEASANT

(PHASIANUS COLCHICUS)

Originally introduced into Britain from Asia in the eleventh century, the pheasant is now a very common sight throughout the country. The bird's natural instinct is to run rather than fly, so during a shoot gangs of 'beaters' encourage the birds to leave the ground. The pheasant nests in a hollow in the ground lined with grass and concealed by vegetation. A clutch of seven to fifteen pale buff eggs, mottled with brown, is laid in April or May. The male may have more than one partner and the females will all lay in the same nest.

SIZE Male 75–90cm. Female 50–65cm.

DESCRIPTION Male has red face, a glossy green head and neck, flecked russet plumage and a long tail. Female is a mottled buff colour with a shorter tail.

HABITAT Farmland, plantations, woods.

VOICE Male makes chattering 'cork-cork' sound.

FLIGHT Explosive take-off, then bursts of flapping followed by long glides.

SIMILAR SPECIES Female similar to black grouse, which has a forked tail.

QUAIL (COTURNIX COTURNIX)

SIZE Up to 18cm.
DESCRIPTION Brown and white streaked pattern on back, broad buff stripes on crown. Female is duller in colour.
HABITAT Farmland, fields. Summer visitor.
VOICE: 'Wick, wick-wick' call.
FLIGHT Rarely flies and only for short distances as a last resort.
SIMILAR SPECIES: Partridge, which is a very similar shape but larger.

The numbers of quail in Britain vary from year to year although the reason is not known. It is a shy bird that is more likely to be heard than seen. However, the sound cannot be relied on as a way of finding it, as it seems to 'throw' its call like a ventriloquist. It eats insects, green shoots and seeds from any type of weed. It nests on the ground, concealing the nest with vegetation, and about twelve to sixteen eggs are laid between April and July. A male often has two partners, both of whom lay in the same nest. The creamy-yellow eggs, blotched with brown, incubate for three weeks; the chicks are able to fly after nineteen days.

(GALLINULA CHLOROPUS) MOORHEN

The moorhen's diet mainly consists of water plants, insects and various invertebrates. It has very long toes, which allow it to spread its body weight and walk on water plants. The toes are not webbed, which makes swimming difficult and the moorhen's style is characterised by the head jerking forward as it moves. It is happy beneath the water and will stay submerged except for the bill if under threat. Its nest is made from a platform of dried water plants built amongst waterside vegetation. It is lined with leaves and five to eleven pale buff eggs, spotted with brown, are laid in April or May. More than one female may share the same nest. At breeding time, this normally shy bird will become very territorial and aggressive towards intruders.

SIZE 32–35cm.
DESCRIPTION Black and dark brown plumage. Bill has red base and yellow tip, white undertail. Green legs.
HABITAT Fresh water.
VOICE Loud clucking and quacking; ringing-type calls.
FLIGHT Flies low over water with legs dangling.
SIMILAR SPECIES Coot, which has a white bill.

COOT (FULICA ATRA)

SIZE Up to 38cm.
DESCRIPTION White bill and frontal shield. Black plumage and lobed webbing to toes.
HABITAT Lakes, rivers, reservoirs.
VOICE 'Kowk' sound made singly or repeated.
FLIGHT Flies low over water with legs trailing, shows white edges to wings.
SIMILAR SPECIES Moorhen, which has yellow-tipped bill with a red base.

Although similar to the moorhen, the coot tends to frequent larger expanses of water. It webbed toes make it a more confident swimmer than the moorhen, and it will dive into the water for food, bringing it to the surface to eat. Its main diet consists of aquatic plants, freshwater shellfish and worms. The male is very territorial, fighting if necessary; as a result, the coot usually needs a much larger territory than the moorhen. Its nest is built in shallow water and is made of reeds. From six to nine stone-coloured, spotted eggs are laid and the chicks are independent after eight weeks.

(VANELLUS VANELLUS) LAPWING

The male lapwing has a very distinctive flight when defending its territory. It will climb up from the ground then go into a twisting, rolling dive, ending with an upward twist accompanied by rapid wing beats. The lapwing eats wireworms, leatherjackets, slugs and snails making it very useful to farmers. It nests in a depression in the ground where four olive and heavily blotched eggs, very pointed at one end, are laid. The chicks are able to run as soon as they are hatched; they often still have pieces of shell stuck to them. In winter lapwings group together and the flocks move to slightly warmer climates such as southern England, Ireland or southern Europe.

Size 30cm.

Description Green upper parts that are tinged with purple with an iridescent sheen. White underparts and a chestnut undertail. White head has a black face patch, a black cap and a long crest.

Habitat Farmland, pasture, mudflats, seashore, fields, marshes.

Voice 'Pee-wit'.

Flight Floppy, showing black and white rounded wings.

GOLDEN PLOVER (PLUVIALIS APRICARIA)

SIZE Up to 28cm.

DESCRIPTION Gold-speckled upper parts with pale underparts that become black in the summer.

HABITAT Spends summers in high moors and moves to lower areas in the winter.

VOICE Melancholy fluting whistle, making 'tloo-ee' noises.

FLIGHT Males display flight has slow, shallow wing action accompanied by song.

SIMILAR SPECIES Grey plover, which has pale grey upper parts.

The golden plover forages for food on the ground or on mudflats, eating either marine invertebrates or worms, insects, slugs and snails. In winter, migrants from Iceland and Scandinavia join the British population. The nest is in a small scrape on the ground and has virtually no nesting material. Usually four heavily blotched, buff eggs are laid at the end of April at two-day intervals. After the third egg is laid the female begins to incubate them and does so for four weeks while the male guards the nest. The chicks are nurtured by both parents for four weeks and are then independent.

(GALLINAGO GALLINAGO) SNIPE

The snipe's main diet consists of worms and invertebrates. Its unique bill greatly assists the bird in its search for food as it digs into the ground. The courtship display involves great climbs assisted by rapid wing beats and then dives with the outer tail feathers spread. The vibrating feathers make a bleating noise. The snipe's well-hidden nest is sited on the ground and the clutch usually consists of four greenish-brown eggs blotched with grey and brown. They are incubated by the female and the young are fed by both parents. The chicks can fly at three weeks.

SIZE 27cm.
DESCRIPTION Light and dark stripes along length of head and upper parts. Very long pointed bill, which is a third of the bird's length. Tip is flexible and very sensitive.
HABITAT Found near inland water. Widespread in Britain.
VOICE Harsh 'scarp'.
FLIGHT Swift and zig-zagging. Soars and dives during display flight.
SIMILAR SPECIES Jack snipe, which is smaller.

BLACK-TAILED GODWIT (LIMOSA LIMOSA)

SIZE 38–41cm.

DESCRIPTION Brown and chestnut
mottled plumage on back. Head
and neck light russet. Whitish
undersides (summer). Grey-brown
upper parts and whiter undersides
(winter). Long legs and bill.

HABITAT Found around the British
coastline in winter. Breeds on damp
marshes and meadows.

VOICE Very noisy 'wick-a, wick-a' call.

FLIGHT Shows black and white tail
and broad wing bar.

SIMILAR SPECIES Bar-tailed godwit,
which has russet undersides.

The black-tailed godwit almost disappeared
from the British Isles in the 1950s when
there were only four breeding pairs.
Fortunately, due to conservation
programmes, their numbers have slowly
increased; numbers of winter migrants have
also risen. In its characteristic display flight,
the male flies noisily into the air with
rapidly beating wings. It twists from side to
side and then glides quietly down, finishing
with a steep dive to the ground. About four
eggs are laid in a dip amongst thick grass
and are incubated by both parents. The
young fly at four weeks.

(NUMENIUS ARQUATA) CURLEW

The curlew arrives at its breeding grounds in early spring, where the male circles to establish his territory. He rises into the air sharply, with very fast-beating wings, and as he goes higher his call also rises in tone and speed. It then dies away as he drops back down. The bird's long beak is used to probe into the ground for the shellfish, worms and insects that make up its main diet. Its nest is made in a shallow hole, normally among clumps of rushes or grass. Three or four buff-coloured eggs with brown blotches are laid, and are mainly incubated by the female. The chicks leave the nest with their parents soon after hatching and fly by the time they are six weeks old.

SIZE 55cm.

DESCRIPTION A sandy, buff colour with white and brown streaks above; undersides are lighter. Very long beak that curls downwards. Long blue-grey legs.

HABITAT Mudflats, moors, marshes, sandy bays. Some winter migrants.

VOICE Makes a 'curlwee' sound and, in the breeding season, a bubbling trill can be heard as it marks out its territory.

FLIGHT Shows white rump and barred buff tail.

SIMILAR SPECIES Whimbrel, which is smaller and generally only found in the Orkney and Shetland Islands between May and September.

(LARUS ARGENTATUS) HERRING GULL

Despite the name, a herring gull actually eats anything including shellfish, small mammals, birds and rubbish. It drops food such as mussels from a height in order to break the shells open. Its population has continued to increase over the years due to the availability of food from refuse dumps. It now nests on buildings as well as cliffs, causing damage and intolerable noise at times. Colonies have sometimes been culled if they have overtaken other bird colonies, such as those of terns and puffins. The herring gull's nest is made from seaweed and grass; two or three olive eggs with brown blotches are laid in May. Immature birds in their first year are brownish and have speckled tails ending in a dark band.

Size 55–67cm. Wingspan 130cm.

Description Pale silvery back with white underparts. Black tip to tail and pink legs. Yellow bill with red spot on lower bill.

Habitat Found all around the coast but moves inland in the winter.

Voice Harsh cries, often making 'qua-qua-qua' call. Also other mewing and wailing noises.

Flight Very strong, graceful, gliding flight using the winds to assist it.

Similar species Glaucous gull, which only visits Britain in very cold winters.

ARCTIC TERN
(STERNA PARADISAEA)

The Arctic tern is the most common of all the terns that breed in Britain. It often mixes with the common tern, making it very difficult to assess its population, but there are estimated to be 14,000 breeding pairs. It winters in the southern oceans and arrives in the UK in May; it leaves Britain in September. The nest is built in a colony which can contain thousands of terns, and the bird uses a shallow scrape on a beach or on ground near the sea. This is lined with broken shells and the clutch normally consists of two eggs. The young are able to swim after only two days, flying about five weeks later.

SIZE 33–35cm.

DESCRIPTION White plumage with black cap and a red bill. Wings are grey and translucent at the tip. Long streamers on the tail and short red legs.

HABITAT Coastal areas mainly in Scotland, the Scottish islands and Ireland.

VOICE 'Keee-ah' call which is short and sharp with a similar emphasis on both syllables.

FLIGHT Primary feathers seem translucent in flight which gives the appearance of a luminous white patch.

SIMILAR SPECIES Common tern, which is almost identical except for a black tip to the bill and a shorter tail.

GUILLEMOT (URIA AALGE)

SIZE Up to 42cm.

DESCRIPTION Dark brown upper parts with white below, streaked flanks (summer). Pale throat and breast with black line behind eye (winter). Dagger-shaped bill.

HABITAT Present all round the British coast in winter. Tends to breed mostly in the north and west.

VOICE 'Cooing' and 'mooin' noises in breeding season; otherwise mostly silent.

FLIGHT Flies low over the sea showing white trailing wing edge.

SIMILAR SPECIES Razorbill, which is blacker with a hooked bill.

The guillemot is a member of the auk family and breeds in colonies numbering up to 70,000 birds. These colonies are located on rocky coastlines, and are near food supplies but protected from potential predators. The guillemot feeds on fish and is able to dive and swim under water to catch them. The birds are often so tightly packed in their colonies that there is only standing room; any available ledge is used. One egg is laid on the ledge in late May. Both parents share incubation by covering it with their belly plumage. The egg is pear shaped to avoid it rolling off the ledge. The chick remains on the nesting site for up to five weeks and then enters the sea; it is another three weeks before it can fly.

(FRATERCULA ARCTICA) PUFFIN

The puffin spends most of its life at sea although, on occasions, it can be blown inland by storms. It is an excellent diver and swims underwater gathering fish, able to carry up to ten sideways in its beak. The puffin breeds in colonies, which number thousands of pairs, and nests in a shallow burrow. Sometimes it digs its own but it often takes them over from rabbits or shearwaters. One egg is laid in June and is incubated by both parents. They feed the chick for forty days but then leave it to fend for itself. It stays in the burrow for up to ten days and, once its plumage has developed, travels to the sea under the cover of darkness, thus avoiding predators such as gulls. In the winter puffins leave the breeding colonies and remain at sea, usually scattered around the North Atlantic.

SIZE 26–29cm.

DESCRIPTION Large, multicoloured bill. Black forehead, cap and nape with white face. White underparts and black back. Orange legs with webbed feet.

HABITAT Remote islands and cliffs around Britain. Chiefly found in the north and west.

VOICE Low growls can be heard in the breeding season.

FLIGHT Rapid whirring flight, low over the sea.

(CUCULUS CANORUS) CUCKOO

The cuckoo is a summer visitor to Britain that arrives from Africa in April and is notorious for using other birds' nests. They return to Africa in July or August. The cuckoo population has decreased in the last six decades, probably due to loss of habitat and climate change. It is an insectivore, eating various types of caterpillars, even the unpalatable ones avoided by most birds. In May, while the males are displaying, the female flies around a chosen territory looking for a nest to use. She eventually selects one already occupied by a pair of birds from a different species, often preferring the nests of the reed warbler, dunnock or pipit. After removing one egg from the chosen nest, she lays a single egg of her own. She may lay as many as twelve, all in different nests and sometimes the egg laid will closely resemble those already present. When the egg hatches, the young cuckoo will push the hosts' original eggs out of the nest and claim all the food provided by the unsuspecting 'parents'.

SIZE 32–34cm.
DESCRIPTION Grey upper parts with light grey underparts barred with dark grey. Long blackish tail, which is rounded.
HABITAT Gardens, woods, heaths and moors.
VOICE Characteristic 'Cuck-oo'.
FLIGHT Wings are scythe-shaped and have shallow beat.

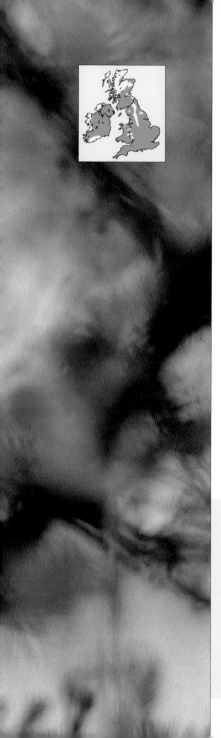

BARN OWL
(TYTO ALBA)

The barn owl is a nocturnal bird that preys mainly on mice, rats and voles, returning to the nest with its victims. It is found throughout Britain but tends to avoid colder, more mountainous areas. It lacks the camouflage of other owls and so tends to hide in holes and old buildings during the day. It often returns to the same roosts and breeding sites each year, favouring places such as old barns, church towers and ruins. Natural holes in trees and cliffs are also used. The barn owl begins to breed as early as February; eggs are laid on a suitable ledge. The average clutch contains four to seven eggs, which are incubated for up to thirty-four days by the female, with the male bringing her food. Occasionally there may be two broods. The young chicks fly at eight to ten weeks.

SIZE 33–35cm.

DESCRIPTION Plumage is a pale sandy brown, flecked with white, grey and brown. Underparts are white. Heart-shaped, white facial disc with large brown eyes.

HABITAT Woods, farmland, villages.

VOICE Snoring sound when roosting but may shriek at other times, giving it the alternative name of screech owl.

FLIGHT Long wings. Appears white when flying at night.

SIMILAR SPECIES Tawny owl, which is much larger with darker plumage.

TAWNY OWL (STRIX ALUCO)

Size 37–39cm.

Description Rich brown plumage with light and dark streaks and bars. Grey-buff circular facial disc with black border. Large dark eyes.

Habitat Woodlands, farmland, parks, gardens.

Voice Characteristic 'to-whit-to-whoo'.

Flight Rounded wings, fluttering wingbeats.

Similar species Barn owl, which is smaller with paler plumage.

The commonest of all British owls, the tawny owl is widespread across a variety of habitats. It is a nocturnal bird; well camouflaged, it spends the days resting in trees. In flight it is totally silent partly due to the structure of the edges of the wings and partly due to its soft plumage. This, combined with excellent eyesight and hearing, makes it a very adept hunter. Its diet mainly consists of small mammals, birds, frogs and fish, which it will grasp in its very sharp talons. Prey is often swallowed whole; regurgitated pellets are then produced containing the indigestible remains. It tends to nest in tree hollows but occasionally uses other birds' old nests. A clutch of eggs contains two to four on average, which are laid at weekly intervals. The young then hatch in turn so there is a large range in the age of the chicks.

(APUS APUS) SWIFT

The swift is a summer visitor from Africa, arriving at the beginning of May and returning in August. It is superb in flight, due to the shape of the wings and the exceptionally long primary feathers, reaching speeds of over sixty miles per hour and heights of three thousand metres. It spends most of its time in the air and feeds exclusively on airborne insects, which it catches using its wide mouth, and it even catnaps on the wing. Swifts can often be seen in groups, excitedly circling and screaming. A young bird will spend the first two years flying until it is ready to breed; contact with the nest hole is its first experience of a solid surface since it fledged. The swift is able to cling to the sides of houses, where nests are normally sited under roofs or in crevices of walls. Two or three eggs are laid; the parents then gather food for the young by storing supplies in their throat pouches before returning to the nest. If food is short the young can go into a torpid state to conserve energy.

SIZE 16–17cm.
DESCRIPTION Black plumage with lighter patch on the chin. Sickle-shaped wings.
HABITAT Breeds in urban areas.
VOICE Shrill, high-pitched screams.
FLIGHT Fast, flickering flight.
SIMILAR SPECIES Swallow, which is larger with long tail streamers.

KINGFISHER
(ALCEDO ATTHIS)

The kingfisher is a shy bird that is very rarely seen; if spotted, it appears as a sudden flash of blue. It is present all over the British Isles except for the far north of Northern Ireland and Scotland. It feeds on tadpoles, small fish, insects and molluscs and will usually dive straight into the water from a perch or plunge from a hovering position to snatch the unsuspecting victim. Fish are killed by beating them against a tree; they are then swallowed whole, head first. The male and female live in separate territories during the winter but pair up by February. Once the weather becomes warmer they select a nesting site, which is often on the bank of a stream. A tunnel up to one metre long is dug, with a chamber at the end where an average of six to seven eggs can be laid. Both parents incubate the eggs and the young emerge after three weeks. They stay in the chamber for four weeks, which means that the tunnel gets very dirty with their droppings. Their parents feed them until they can fish for themselves.

SIZE 15–17cm.
DESCRIPTION Dark blue crown with lighter blue flecks, bright blue-green upper parts, chestnut below. Long, dagger-like, black beak. Red feet and legs.
HABITAT Rivers, streams and lakes. Coastal areas in winter.
VOICE A shrill 'cheet'.
FLIGHT Fast flight low over the water, with whirring wings.

GREEN WOODPECKER
(PICUS VIRIDIS)

SIZE 28–33cm.

DESCRIPTION Green plumage above with paler green underparts. A yellow rump, crimson crown and cheek patches; black eye patch.

HABITAT Deciduous woodland, open grassland, heaths, gardens.

VOICE A laughing 'yah-yah-yah'.

FLIGHT Deep and undulating, showing the yellow rump.

SIMILAR SPECIES Great spotted woodpecker, which has black, white and red markings.

Unlike other woodpeckers, the green woodpecker hardly ever drums against trees. It predominantly feeds on the ground, using its beak to probe for beetles, ants, flies and wood-boring insects, then licking them up with its long tongue. During the courtship display the male spirals around tree branches pursuing the female. When two males are fighting over a female, they spread their wings and tail feathers, raise their crests and sway their heads from side to side. The green woodpecker nests in trees, usually laying five to seven eggs in a bare chamber; the clutch is incubated by both parents for nineteen days. The young are fed for three weeks by their parents with a regurgitated milky substance which is formed from the insects they have eaten.

GREAT SPOTTED WOODPECKER
(DENDROCOPOS MAJOR)

A drumming noise gives away the presence of the great spotted woodpecker. It is made by a rapid succession of blows on a wooden surface such as a tree or telegraph pole. This is partly done to mark out territory but also to source food under the bark. It eats mainly insects and their larvae but when these are unavailable it also consumes berries and nuts. During the courtship display the male pursues the female by twisting around branches. Both birds then create a nest by excavating a hole in a tree. Up to eight eggs are laid in the bottom of the hole, and these are generally incubated by the female for up to sixteen days. Both parents feed the chicks.

SIZE 22–23cm.

DESCRIPTION Black and white head and upper parts. Creamy-coloured undersides and red flashes on back of head and undertail.

HABITAT Any habitat where there are suitable mature trees with some dead branches.

VOICE Drumming and 'kek' or 'chack' sounds.

FLIGHT Undulating pattern. Shows white wing patches and barring across back.

SIMILAR SPECIES Lesser spotted woodpecker, which is smaller.

SWALLOW (HIRUNDO RUSTICA)

Size Up to 19cm.

Description: Black-blue upper parts with band around neck. Cream undersides. Male has a russet-coloured throat with long tail streamers, the female has a shorter tail and is duller in colour.

Habitat Towns and villages.

Voice Long musical twitters and a curt 'chirrup' call.

Flight Fast, darting and swooping.

Similar species Swift, which is smaller with a shorter tail.

The swallow spends much of its time in flight feeding on airborne insects; it swoops down to drink without landing. It is a summer visitor from South Africa that can appear as early as March, with the male arriving first. It normally returns to the same breeding site where a cup-shaped nest is built from mud, grass and feathers, usually on the wall of a barn, porch, shed or garage. The female incubates the first clutch of three to six eggs for about fifteen days. The young are fed for three weeks and leave the nest soon afterwards. The parents then raise a second brood, but have been known to produce three in one season. The birds make the long journey back to South Africa between July and September, often gathering on telephone lines before they leave.

(DELICHON URBICA) HOUSE MARTIN

A summer visitor to Britain, arriving from Africa in April, the house martin was originally a cliff-nesting bird but has now adapted to live alongside man. On arrival, it looks for good food sources and can often be seen by lakes where there is an abundant supply of insects; it is an aerial bird that feeds on the wing. The nest is built of mud and usually takes two weeks to complete. Pairs will bring in pieces of mud and stick them together to create a quarter-globe structure that can be seen on the outside of a building. It has a very narrow entrance at the top and is lined with feathers and grass. An average of four or five eggs are laid and are then incubated for up to nineteen days. After three and a half weeks the young leave the nest and up to two more broods may then be reared during the rest of the breeding season. The birds leave Britain by the end of October.

SIZE 12cm.

DESCRIPTION Blue-black upper parts, white undersides and rump. Short, shallow-forked tail.

HABITAT Cliffs, bridges, towns and villages.

VOICE Rattling twitter.

FLIGHT Swooping and wheeling. The white band on the rump can be clearly seen.

SIMILAR SPECIES Sand martin, which has brown colouring and a less forked tail.

GREY WAGTAIL (MOTACILLA CINEREA)

SIZE: 18–19cm.

DESCRIPTION Male has grey upper parts, a black throat and yellow undersides, white eye stripe and a long dark tail. Female lacks the black throat and has duller colours.

HABITAT Beside fast-moving water including streams, rivers, rapids.

VOICE High pitched 'chee-seek'. Sometimes trills.

FLIGHT Shows yellow breast and long tail.

SIMILAR SPECIES Yellow wagtail, which has yellow upper parts and is found in damp, marshy areas.

The grey wagtail is a bird constantly on the move as it searches for food, often taking insects in mid-flight. Its preference is for flies, midges, mayflies and small dragonflies. During the courtship display the male fluffs up his plumage and fans his tail feathers as he slowly flies from perch to perch. The nest is built in April; the pair chooses a hollow among tree roots or a ledge on walls by the water. They use grasses and moss to build the nest, then line it with hair. Four to six eggs are laid and are incubated for two weeks. The young are able to fly after a further two and a half weeks. During the winter grey wagtails may move further south, sometimes going to mainland Europe and returning in the spring.

(MOTACILLA ALBA YARRELLII) PIED WAGTAIL

The pied wagtail spends most of its time near any source of water. It is equally happy anywhere with a good food supply where it can eat flies and other insects. During the courtship period more than one male is likely to pursue a female and they each perform an undulating display flight. The female builds a nest in a crevice or a hole in a building, tree, cliff or bank. The nest is made from twigs, grass and dead leaves, and is lined with feathers and wool. A clutch contains an average of five or six eggs and after two weeks of incubation both parents feed the chicks. Evidence shows that some of the British population is migratory, with birds travelling as far south as Morocco.

SIZE 18cm.

DESCRIPTION White underparts and eye patches. Cap, back and breast are black. Long black tail.

HABITAT Urban areas, open land, farmland. Often near water.

VOICE Twittering call with 'chis-ick' sound.

FLIGHT Deeply undulating.

SIMILAR SPECIES Grey wagtail, which has yellow undersides.

SKYLARK (ALAUDA ARVENSIS)

SIZE 18–19cm.

DESCRIPTION Brown plumage with white outer tail. Small crest.

HABITAT Grassland, moors, cultivated land, meadows.

VOICE Very musical song made as it flies. Also a 'chirrup' sound.

FLIGHT Hovers or circles high above the ground.

SIMILAR SPECIES Woodlark, which is smaller with more richly coloured markings.

The skylark is a bird that has significantly decreased in numbers over the last thirty years, although the reasons for this decline are not clear. It favours rural habitats and can be found all over the British Isles. It maintains its song for several minutes as it rises several hundred feet before eventually sinking back to the ground. The nest is sited on the ground and made from dried grass. Three to five eggs, covered with brown spots, are laid and the young fly after three weeks. Winter migrants come to Britain from northern Europe.

ROBIN
(ERITHACUS RUBECULA)

Robins living close to man are extremely tame and will happily come into a freshly dug garden to look for worms, even if there are people there. However, those that inhabit rural areas are more wary and away from Britain the robin is renowned for being particularly shy. The birds pair up as early as January when the male will sing loudly, protecting his territory. The sexes are identical and the male, unable to tell whether an intruder is male or female, automatically behaves aggressively. Only the female will persist in her approach until the two finally make a bond; other males will fight for the territory or retreat. It is not until the weather improves that the female begins to build a nest with moss, dead leaves and hair, using crevices in hollows or trees or inside outbuildings. At this point the male begins to feed the female and will continue to do so while she incubates the eggs; a clutch normally contains five or six, with one laid each day. The female loses plumage from her breast and the blood vessels enlarge to allow greater heat transference to the eggs. They hatch after a fortnight and the chicks leave the nest after a further two weeks.

SIZE 14cm.

DESCRIPTION Brown crown and back. Red breast and face with white undersides.

HABITAT Widespread throughout the British Isles.

VOICE High-pitched, warbling song and 'tick-tick-tick' call.

FLIGHT Red and white undersides clearly visible.

SIMILAR SPECIES Red-breasted fly catcher, where the red markings are absent from the face.

DUNNOCK (PRUNELLA MODULARIS)

SIZE 14–15cm.
DESCRIPTION Grey head and undersides, dull brown with speckled flanks. Thin beak.
HABITAT Woodland, gardens, farmland, parks, coasts.
VOICE High-pitched, repeated song and shrill 'seek' call.
FLIGHT Rapid flight; wings appear rounded.
SIMILAR SPECIES House sparrow, which has more uniform grey undersides and a stouter bill.

The dunnock is also known as the hedge sparrow but is actually a songbird, and is found throughout Britain. It often falls victim to the cuckoo, which will lay its own eggs in the dunnock's nest. The newly-hatched cuckoo pushes the original eggs out and the unsuspecting new 'parents' end up rearing the cuckoo rather than their own young. The dunnnock's nest is sited in hedges or bushes and is built by both the male and female. Plant stems, twigs and moss are used and it is then carefully lined with hair and wool. A clutch normally contains three to six bright blue eggs that are incubated by the female, though she leaves the nest to feed when she needs to do so. The eggs hatch after two weeks; the young are fed by both parents and fly after a further fortnight. Two or three broods are normally raised in a season.

BLACKBIRD
(TURDUS MERULA)

The blackbird is one of the most common birds in Britain. It eats worms and insects, also surviving on fruits and berries when in season, and will make occasional visits to bird tables. It can frequently be seen turning over dead leaves beneath shrubs and trees, searching for food. The nest is mainly built by the female, using dried grass, twigs and moss; it is then lined with mud. Usually three eggs are laid and are incubated for two weeks. The young leave the nest two weeks later but still depend on their parents for food for a further three weeks.

SIZE 24–25cm.

DESCRIPTION Black plumage with yellow eye ring and orange bill (male), brownish plumage with blurred spots on breast (female).

HABITAT Widespread throughout the British Isles.

VOICE Fluting varied song with 'chack' call.

FLIGHT Rapid flight with flickering wings, sometimes glides and flicks tail up on landing.

SIMILAR SPECIES Ring ouzel, which has a white crescent at throat.

SONG THRUSH (TURDUS PHILOMELOS)

SIZE 23cm.

DESCRIPTION Brown head and upper parts. White undersides with brown spots.

HABITAT Woodland, parks, gardens, farmland. Widespread throughout the British Isles.

VOICE A thin sounding 'seep' call and song consisting of a repeated series of notes lasting as long as five minutes.

FLIGHT Orange-yellow underwings and uniform brown tail are visible.

SIMILAR SPECIES Mistle thrush, which is slightly larger with greyer plumage on head and back.

The song thrush tends to sing from a high perch such as a chimney pot or television aerial and can be heard on fine days throughout the year. Its diet mainly consists of worms, slugs, insects and snails; if necessary, the thrush smashes shells against a rock in order to break them and access the contents. During the breeding season, the female builds a nest using grass and twigs, then lines it with mud. Four to six eggs are laid and are incubated for two weeks. The young birds then spend up to sixteen days being fed by both parents before they leave the nest.

(SYLVIA BORIN) GARDEN WARBLER

Arriving from tropical and southern Africa in April, the garden warbler is renowned for its melodic song. Its main diet consists of insects but prior to the migration back to Africa in September it will feed on as many berries as possible, with its beak often becoming stained with the juice. It likes thickets, brambles and hedges where it can build nests with plenty of cover and so tends not to frequent gardens unless they are overgrown. The nest is built by both birds from twigs and grass stems, and is then lined with hair and roots. A clutch of four or five eggs is laid. These hatch after only twelve days and it is only a further ten before the chicks leave the nest. The garden warbler's nest is built very close to the ground making them vulnerable to predators, so the young need to be independent as soon as possible to maximise their chances of survival.

SIZE 14cm.

DESCRIPTION Brown and buff indistinct plumage. Short beak and bluish legs.

HABITAT Woodlands, hedgerows.

VOICE Extended melodious warbling song and sharp 'tack' call.

FLIGHT During courtship flight, male will spread his tail and flutter his wings as he turns towards the female.

SIMILAR SPECIES Blackcap. Male has black cap, which is brown in the female.

CHIFFCHAFF (PHYLLOSCOPUS COLLYBITA)

Size 10–11cm.

Description Grey-green upper parts and pale yellow below. An indistinct eye stripe and black legs.

Habitat Areas where there are tall trees and bushy undergrowth. Not found in northern Scotland.

Voice Two notes, a 'chiff, chaff' constantly repeated for about twelve seconds.

Flight In courtship flight the male will appear to 'float' down to the female.

Similar species Willow warbler, which has more olive-coloured plumage with yellower underparts and paler legs.

The chiffchaff comes to Britain from southern Europe or Africa between March and May and returns in October. The male arrives first to establish a territory, often where it was originally raised. The female then builds a nest amongst dense vegetation, near to the ground. It is nearly spherical in shape with a small entrance, and it is made from moss, stalks, bracken and leaves. Feathers are used as lining material and an average clutch of six or seven eggs is laid. After less than a fortnight these hatch, and the young remain in the nest for a further two weeks.

(AEGITHALOS CAUDATUS) LONG-TAILED TIT

The long-tailed tit is a very sociable bird and is almost always seen in groups. It feeds on spiders and insects, supplemented by buds and seeds. Due to their size, they lose heat very quickly and the population can be severely affected in a hard winter. It is renowned for its skill in nest building, weaving an amazing oval-shaped nest using moss combined with cobwebs, wool and hair. This is then lined with a vast number of feathers. It is sited in a tree fork or in a shrub and a clutch of eight to twelve eggs is laid. The eggs are incubated for about sixteen days and the young chicks then spend up to a further three weeks in the nest.

SIZE 14–15cm.

DESCRIPTION White and black upper parts tinged with pink. White head with dark stripes over eyes. White undersides also tinged with pink on the belly. Very long black tail edged with white.

HABITAT Woodland, undergrowth, farmland, hedges.

VOICE Call consists of repeated 'tup' or 'see' notes.

FLIGHT Feeble flight with short, rounded wings. Very long tail conspicuous in flight.

SIMILAR SPECIES Bearded tit, which is predominantly brown with a grey head.

BLUE TIT
(PARUS CAERULEUS)

During summer the blue tit feeds mainly on insects, which it searches for on the tips of shoots and twigs. In winter it also eats nuts and seeds and is a regular visitor to bird tables. It is a very lively and agile bird that can cling onto objects at amazing angles in order to feed. Nests are made in holes in trees or bird boxes, and are filled with dead leaves, wool, moss and hair. A clutch can contain up to fourteen eggs, laid daily, with incubation starting when the clutch is almost complete. The young chicks are usually fed with caterpillars and are able to fly within three weeks.

SIZE 11–12cm.

DESCRIPTION Greenish back and bright blue wings and tail. Yellow undersides. White head with bright blue crown, black eye stripe and black bib.

HABITAT Widespread and numerous, frequenting gardens, woodland and farmland.

VOICE Fast trilling song and 'tsee-tsee-tsee-sit' call.

FLIGHT Yellow underparts clearly visible.

SIMILAR SPECIES Great tit, which has black crown and blue-grey wings.

COAL TIT (PARUS ATER)

SIZE 10–12cm.

DESCRIPTION Olive-grey upper parts, black head and bib with white cheeks and nape. Pale undersides.

HABITAT Woodland environments, especially with coniferous trees, gardens and parks.

VOICE Call is a high-pitched 'tseet'. Repetitive song making 'wheat-see, wheat-see' notes.

FLIGHT Double white wing bar is clearly visible.

SIMILAR SPECIES Marsh tit, which has brown upper parts.

The smallest of all British tits, the coal tit feeds on insects such as beetles, moths, spiders and flies. Seeds supplement this diet, together with any scraps from bird tables. The coal tit survives well in cold weather as it feeds on insects living underneath bark. It is a sociable bird and tends to feed with other members of the tit family. Nests are sited in holes, usually in trees or walls, and are built by both birds. A neat cup shape is made from moss, hair and feathers with the female incubating the clutch of seven to ten eggs for two weeks, while being fed by the male. The young remain in the nest for up to nineteen days and are independent after a further two weeks.

(SITTA EUROPAEA) NUTHATCH

The nuthatch eats acorns, seeds, beech nuts and hazelnuts and anchors these in the crevices of tree bark, using its sharp bill to split them open and access the kernels. Using its feet, it is able to hop up and down tree trunks with great agility. The nuthatch stays within a mile or two of where it was born and uses nest boxes or holes in trees or walls to breed. Any cracks in the hole are filled with mud and the bottom is lined with bark or dried leaves, later used to cover the eggs. These are laid daily and an average clutch of six to nine eggs is eventually produced. They are incubated by the female for two weeks and the young birds fly after about three weeks.

SIZE 13–14cm.

DESCRIPTION Blue-grey upper parts, buff undersides with reddish-coloured flanks. Short tail, strong, sharp bill and prominent black eye stripe.

HABITAT Woods, parks and large gardens.

VOICE 'Chwit' call and 'toowee, toowee' song.

FLIGHT Silvery-white and grey under wings with black patch at 'wrist' and black v-stripe on tail are clearly visible.

TREECREEPER (CERTHIA FAMILIARIS)

SIZE 12–13cm.

DESCRIPTION Brown upper parts with pale undersides and a white eye stripe. Bill is long with a downward curve. Large claws and long tail.

HABITAT Deciduous woodland.

VOICE 'Tseeu' call and thin, high-pitched 'toowee' song.

FLIGHT Undulating flight showing the long tail and multiple wing bars.

SIMILAR SPECIES Short-toed treecreeper, which has light brown flanks.

Its long tail and large claws enable the treecreeper to make its jerky movements up tree trunks in search of food. It hops with both feet while the tail is used as a prop. It cannot, however, descend in this fashion, so it needs to fly to begin its journey up another trunk. Its diet consists of insects such as beetles, woodlice, spiders and earwigs. Nests are sited on trees, behind loose bark or in cracks in the trunk. They are built from roots, dried grass and moss on a base of twigs and are then lined with wool and bark fragments. A clutch usually consists of about six eggs and the young are fed by both parents for two weeks.

(GARRULUS GLANDARIUS) JAY

The jay is a member of the crow family although much more brightly coloured than its relatives. Its favourite habitat is woodland with plenty of oaks where it tends to feed on the ground looking for acorns, beech nuts and hazelnuts. They will also eat soft fruit, such as pears or cherries. In the summer months eggs belonging to other birds will be sought out and they have also been known to eat insects, mice, fish and bread; they do not, however, eat grain or carrion. In the autumn acorns are hoarded and several thousand are stored to provide a constant food supply. The spring courtship display culminates in the male turning sideways to the female and raising its crest and body feathers. The nest is sited in a tree, and is made with twigs, earth and hair. A clutch of five to seven eggs is incubated for two weeks with the young birds spending a further three weeks in the nest.

SIZE 33–35cm.

DESCRIPTION Buff plumage, white rump and a black tail. Black wings with white wing bar and a blue wing patch. Streaked crest and black 'moustache' stripes.

HABITAT Farmland, woods, gardens, parkland.

VOICE Very noisy 'skaaark' call.

FLIGHT Laboured flight showing rounded fingered wings with the white tail patch clearly visible.

JACKDAW (CORVUS MONEDULA)

SIZE 32–34cm.

DESCRIPTION Black plumage with slate grey markings to crown, nape and upper breast. White eyes.

HABITAT Parks, open woodland, farmland.

VOICE Metallic sounding 'chack' call.

FLIGHT Direct flight with rapid wing beats showing pointed wings.

SIMILAR SPECIES Rook, which is larger and has longer, paler bill. Plumage is all black.

The main diet of the jackdaw consists of seeds, insects, carrion, fruit and, occasionally, young birds or eggs. For most of the year, jackdaws live in mixed flocks. At the start of the breeding season they pair and the pairs generally nest in colonies. Nests are built on chimneys or in holes, using varying amounts of nesting material according to the chosen site. Other birds' nests may also be used. Up to six eggs are laid in late April and the female incubates these for about seventeen days. The young leave the nest after a further four weeks.

(PASSER DOMESTICUS) HOUSE SPARROW

The house sparrow has always lived alongside man, needing to do so for shelter and food. It lives in flocks which generally form in the late summer and usually roost together. It eats mainly seeds, grain and insects, sourcing these from gardens and farms. It also happily eats the contents of any bird table. In courtship displays, several chirping males surround a female. She then raises her tail and bill, droops her wings and pecks at them. The house sparrow is a prolific breeder, producing three broods a year with a clutch of up to six eggs laid at each time. The nest is usually sited in a hollow on a building or in dense shrubs. Often there are several close together, making a small colony. The eggs hatch after two weeks and the young birds can fly after a further fortnight.

SIZE 14–15cm.

DESCRIPTION Back is a rich brown with dark brown streaks. Male has a grey crown and black bib. The female has a brown crown with no bib. Double white wing bar.

HABITAT Generally live close to man: farms, towns and cities. Very common and widespread.

VOICE Monotonous song consisting of various chirruping notes. Call is a 'chirrup' sound.

FLIGHT Broad white wing bar is visible.

SIMILAR SPECIES Dunnock or hedge sparrow, which has a grey head and chest, and a thinner beak.

BULLFINCH
(PYRRHULA PYRRHULA)

The bullfinch devours the buds from fruit trees and flowering shrubs, potentially causing a great deal of damage to fruit growers. It has been calculated that one bird can eat half the buds from a pear tree in a single day. On these sites growers are legally allowed to shoot the birds. The bullfinch also feeds on seeds and, occasionally, fruit and insects. Nests are sited in trees or bushes and constructed from twigs, roots and hair. The female incubates four or five eggs for two weeks with the male feeding her during this period. Both parents then feed the chicks with regurgitated food, storing large amounts in special cheek pouches so that they do not have to keep returning to the nest. The young fly after about two weeks. The eggs are very vulnerable to predators so two or three broods are raised each year to ensure that some survive.

SIZE 13–15cm.

DESCRIPTION Grey upper parts, a black cap, red underparts and a white rump.

HABITAT Woodland, parks, farmland, gardens. Present throughout the British Isles.

VOICE Call is a whistling 'peeu' sound.

FLIGHT Deeply undulating, hesitant flight. White wing bar is clearly visible.

SIMILAR SPECIES Chaffinch, which has a grey cap and pink-brown undersides.

GOLDFINCH (CARDUELIS CARDUELIS)

Size 12–13cm.

Description Black and white head with red face. Black wings with gold wing bar. White rump and black tail. Whitish beak with black tip.

Habitat Gardens, open woodland, parks and farmland.

Voice 'Dee-dee-lit' call and a warbling, prolonged song.

Flight Broad yellow wing bar and white rump are clearly visible.

Similar species Greenfinch, which is olive-green in colour.

The goldfinch is a very pretty bird that feeds on the seeds of thistles, groundsel and other annual weeds and supplements them with insects. It is unique in its ability to hold a plant with its feet while feeding. Due to its appearance and pleasant song, the goldfinch is often chosen as a caged bird. It will often congregate with others and a flock is known as a 'charm'. The female builds a well-hidden nest among the upper branches of a small tree. She uses plant material to weave a neat cup and lays a clutch containing four to seven blue-white eggs, streaked in red-brown. These hatch after two weeks' incubation and the young chicks are fed with regurgitated food for a further fortnight. After leaving the nest they are still dependent on their parents for food for another seven days.

(EMBERIZA CITRINELLA) YELLOWHAMMER

The yellowhammer delivers its loud song on a high perch from February to the end of August. It feeds on the ground and can be seen in flocks in the winter, often mixing with other buntings and finches. The courtship display is quite manic, with the male chasing the female in a frenzied, twisting flight that can result in both birds tumbling in the trees. Breeding begins at the end of April when a nest is built by the female from moss, grass and plant stems, lined with grass and hair. It is sited on the ground or low down in a hedge. The average clutch contains three or four white eggs, scribbled with a dark purple-brown. These hatch after thirteen days; the young are cared for by both parents and can fly about ten days later.

SIZE 16–17cm.

DESCRIPTION Bright yellow breast and head. Light chestnut above and white tail feathers.

HABITAT Widespread except for the far north-west of Scotland. Farmland, scrub, heaths, allotments – open areas with low cover and few trees.

VOICE Sharp 'twick' call and a loud insistent song consisting of repeated 'zit' notes ending with a 'zeee'.

FLIGHT White tail feathers and chestnut rump can be seen.

SIMILAR SPECIES Cirl bunting, which has a black and yellow head and a green breast band. Olive-green rump is visible in flight.

A FIELD GUIDE TO THE WILDLIFE OF THE BRITISH ISLES

AMPHIBIANS & REPTILES

SMOOTH NEWT
(TRITURUS VULGARIS)

The smooth newt is the most common and widespread newt on the British mainland; it lives anywhere as long as there is water in which it can breed. Its diet mainly consists of slugs, worms and insects, which it hunts for at night. The newt catches prey using its sticky tongue and swallows it whole. During the day it hides in the grass or under stones. It hibernates on land during the winter and emerges in the spring to mate. The newts move to the breeding pool and the eggs are wrapped individually in the leaves of water plants. After the adults have left the pool, the eggs gradually develop into tadpoles and then tiny newts, which move onto land at the end of the summer. They are sexually mature after two years, when they return to the water.

SIZE Up to 10cm, including the tail.
DESCRIPTION Dull, yellow-olive, spotted body. Creamy undersides. Skin is equipped with sensory cells. Female duller in colour than the male. During breeding season male develops a high crest along its back and bright orange undersides.
HABITAT Streams and ponds, open woodland, gardens.
SIMILAR SPECIES Palmate newt, which is smaller.

GREAT CRESTED NEWT
(TRITURUS CRISTATUS)

The loss of breeding ponds has led to a substantial reduction in numbers of the great crested newt. Conservation groups are currently reintroducing them into specific sites. This newt mainly feeds on insects, worms, tadpoles, slugs and snails. It does not always hibernate, instead remaining in the water; those that do hibernate find a crack in the ground or a place under logs. In March the hibernating newts return to the water and the female lays two to three hundred eggs in April. She may continue to do this until July when the adults leave the water. The final development of the offspring may take place in the winter or the following spring.

SIZE Body length, including tail, up to 15cm. Female is larger than the male.

DESCRIPTION A slimy, warty skin with a grey-black spotted body and orange-spotted undersides. Male has a silver-streaked tail and a crest down the centre of the back, which is absorbed at the end of the breeding season.

HABITAT Ponds and small lakes. Numbers have declined and it is now a protected species.

SIMILAR SPECIES Smooth newt, which is smaller with a smoother skin.

COMMON FROG
(RANA TEMPORARIA)

The common frog is a very widespread and abundant amphibian that can cope in very cold conditions. It eats insects, slugs and snails, moves only by hopping and can cover up to half a metre in a single jump. It hibernates from October to February in a sheltered place on land or at the bottom of a pond. The common frog is sexually mature at three years and migrates back to the same breeding pond each year. During the breeding season the male develops a larger, black, rough patch on its first finger to enable him to grasp the female; he then attracts a mate by croaking. The female lays between one and two thousand eggs, surrounded by transparent jelly, at the bottom of the pond. These then fill with water and rise up to the surface, where they float. Tadpoles hatch after about two to four weeks. By the time they are three months old they have grown arms and legs and are adapted to living on land. The tail is absorbed and by the beginning of June the young frogs are ready to leave the water.

SIZE Up to 9cm. Male usually smaller than female.
DESCRIPTION Colours vary from brownish to olive green with dark-coloured blotches. There is always a dark triangular patch behind the eyes. Has smooth skin and fully webbed hind feet.
HABITAT Likes wet habitats such as ponds, open woods. Needs slow-moving water.
SOUNDS Croaking.
SIMILAR SPECIES Marsh frog, which is larger with a more pointed snout.

COMMON TOAD (BUFO BUFO)

SIZE Up to 13cm. Male is smaller than female.

DESCRIPTION Dry, warty skin, olive brown in colour

HABITAT Marshes, moors and woodlands. Goes to ponds and lakes to breed. Widespread throughout England and Scotland.

SOUNDS Croaking (only the male).

SIMILAR SPECIES Natterjack toad, which has a yellow line down the middle of its back.

A toad moves by walking, distinguishing it from a frog which leaps. If threatened, it secretes a strong-smelling, poisonous substance from glands on its back. An evening hunter, it catches prey such as insects and small animals by using its long, sticky tongue. It hibernates in dry places, such as under logs, from October to March. When it emerges from hibernation, it travels to the breeding pond; unfortunately many are killed as they cross roads en route. The male develops special grasping pads on his palms at pairing time with which he clasps the female from behind and fertilises the eggs as they emerge. The spawn is laid in long strings up to three metres in length. After spawning the adults leave the pond and live alone. The eggs initially develop into tadpoles, which finally become toads after about fifteen weeks. The young toads leave the water in June or July. The male is fertile at three years, the female at four.

(LACERTA VIVIPARA) COMMON LIZARD

A rapid mover, the common lizard runs with a gliding movement and is also a good swimmer. It sloughs (moults) occasionally but the old skin tends to be scraped off in pieces, making the lizard look ragged at times. It hunts by day but hides at night and in cool weather, basking during daylight to gain heat. Its diet mainly consists of spiders, flies, beetles, moths and caterpillars. Predators of the common lizard include birds, snakes and rats. It is able to shed its tail in times of danger; a new one grows, but not to the same length as the original. The lizard hibernates under stones and in cracks in walls during the winter and emerges in spring ready to mate. The female retains her eggs until they are fully developed; the young are then born in a transparent membrane that breaks immediately. There are usually five to eight in a litter. They are fully independent at birth, eating the same diet as their mother.

SIZE About 15cm long, including the tail.

DESCRIPTION A dull, grey-brown colour with a dark stripe along the back and further blotched markings. Female tends to be paler and fatter. Has sharp-clawed toes that can spread widely to grasp rough vertical surfaces easily.

HABITAT Coastal cliffs, downs, moors, heaths, grassland. Widespread throughout Britain.

SIMILAR SPECIES Sand lizard, which is slightly longer with the male having green flanks.

SLOW WORM (ANGUIS FRAGILIS)

SIZE Up to 50cm.
DESCRIPTION Colours vary from grey to dark brown. Body is same thickness all the way along.
HABITAT Grassland, gardens, dry heaths. Very common throughout Britain, although not often seen.
SIGNS Fragments of shed skin.
SIMILAR SPECIES Smooth snake, which has a more tapered body.

The slow worm is actually a legless lizard, albeit one with a snake-like appearance but, unlike a snake, it has eyelids. It is a slow-moving creature with a diet that mainly consists of slugs. During the day it basks in partial sunshine, choosing the branches of trees, or rests under stones. The slow worm's predators include cats and hedgehogs and it is able to shed its tail to make an escape; this then partly grows back. It hibernates underground from October to March. It mates in April, during which time males can be seen fighting. Up to twelve young are born in late summer; the egg breaks open within seconds of emerging from the female's body and the young are about seventy centimetres long. They are independent from birth.

(CORONELLA AUSTRIACA) SMOOTH SNAKE

One of Britain's rarest animals, the smooth snake – which is harmless to humans – is now a protected species. Its survival is threatened by the loss of suitable habitat. It eats mainly lizards and some small animals, catching prey by winding its body around the victim before swallowing it whole. It spends days basking in the sun although it prefers to stay under a stone or amongst heather. It hibernates during the winter and emerges in the spring to pair and mate. Eggs are retained in the female's body until they are fully developed and ready to hatch. Up to fifteen young are born in August or September and are about fifteen centimetres long. They are fully independent from birth.

SIZE Up to 70cm long.
DESCRIPTION Slender body that is grey-brown with indistinct dark markings along the back.
HABITAT Sandy heaths. Found only in Surrey, Hampshire, Dorset and West Sussex.
SOUNDS Hisses.
SIMILAR SPECIES Adder, which has a distinct zig-zag marking along back.

GRASS SNAKE
(NATRIX NATRIX)

Britain's largest snake is totally harmless. The grass snake is a very strong swimmer that will happily hunt in ponds. Its diet mainly consists of tadpoles, frogs, newts and small fish, and it takes larger prey on land to eat. It feeds in the early morning and then spends the day basking in the sunshine. One meal can satisfy it for ten days. Its skin is sloughed (shed) up to twelve times a year. Hibernation takes place from October to March, often in wall crevices or under tree roots. Mating occurs during April or May and the female then looks for somewhere warm to lay the eggs, such as a compost or manure heap. The eggs are matt white and connected with a string. The young snakes hatch in late summer and are about twenty centimetres long at birth.

SIZE Up to 150cm. Male is smaller than female.
DESCRIPTION Olive green with black bars along the body. Two yellow or white collars around the neck. Often darts out its forked tongue. Pupils of eyes are circular and there are no eyelids.
HABITAT Widespread throughout England and Wales, often near to human habitation. Prefers damp habitats.
SOUNDS Hisses.
SIMILAR SPECIES Adder, which is shorter and fatter with zig-zag markings.

ADDER (VIPERA BERUS)

Also known as the viper, the adder is widespread throughout Britain. It is the only poisonous British snake. Its venom can cause swelling and fever but is rarely fatal to humans; antivenin is available. It is a very shy creature that will normally keep away from people. The adder is active during both day and night and its main diet consists of mice, voles, lizards and frogs. It eats only occasionally; one meal can satisfy it for up to a week. It kills its prey with a bite from the fangs which inject venom at the same time. It then opens its mouth widely, taking advantage of its loosely hinged jaw to swallow the animal whole. Hibernation takes place from October to March and several adders will shelter together in an existing hole. It pairs in April and the young are born in August or September. The eggs are retained until full development has taken place; they then hatch within the mother and are born live and fully independent. The mother spends no time rearing them at all.

SIZE 50–80cm long.

DESCRIPTION Dark zig-zag line along the back and V mark on the back of the head. Males are grey and females are brown with longer and fatter bodies.

HABITAT Sand dunes, moors, heaths, downs, farmlands, woods and hedgerows.

SOUNDS Hisses.

SIGNS Sheds its skin occasionally; this may be found.

SIMILAR SPECIES Smooth snake, which is slimmer with more indistinct markings.

(FORMICA RUFA) WOOD ANT

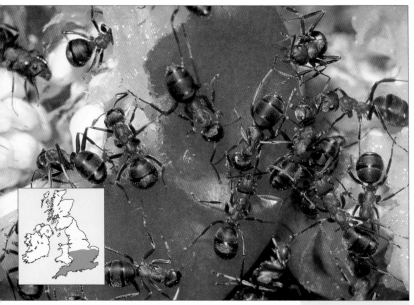

The wood ant is a very sociable insect, building nests in groups to create large communities. Each community has its own hunting territory, which it guards fiercely against others. A dome is built from leaves, twigs and pine needles over the nest to protect it; underneath a series of tunnels hold up to 300,000 workers who spend most of their time finding a supply of food. They hunt for insects such as caterpillars, injecting them with formic acid secreted from a gland in the abdomen; the acid paralyses the prey's internal tissues. The prey is then taken back to the nest and some food is regurgitated to feed the larvae. After mating, the queen begins to lay her eggs using her old nest, or goes to a new one. The workers move the eggs to the warmest part of the nest to aid their development, while the queen continues to lay many thousands each year. After being fed by the workers, the larvae form cocoons, which the workers cut open when the new ants are ready to emerge.

SIZE 7–11mm.

DESCRIPTION Black head and abdomen with brown thorax and legs.

HABITAT Woodland. Most common in the south of England, absent from Scotland.

SIMILAR SPECIES Meadow ant, which is pale yellow in colour.

BLACK BEAN APHID (APHIS FABAE)

SIZE 2mm.
DESCRIPTION Black body, with or without wings.
HABITAT Gardens with suitable plants such as broad beans or nasturtiums.
SIMILAR SPECIES Rose aphid, which is more commonly known as the greenfly. Begins by living on roses then moves to other plants in the summer.

These aphids are also known as blackfly. They cover young shoots in the early summer and suck up the sap. The sap is very high in sugar but contains little protein, so the aphids need very large amounts. The excess sugar is then secreted as honeydew, which provides a source of food for ants. Honeydew will often drip onto plants making a sticky mark that can turn to mould later, blackening the leaves. These aphids also spread viruses from plant to plant. Black bean aphids' predators include ladybirds and birds. They mate in late summer and the females lay eggs which hatch in the following spring. That generation of females then produce summer aphids by parthenogenesis – without the need for a male. The colonies continue to become more crowded and eventually winged females are produced who begin the cycle again.

(ALEYRODES BRASSICAE) CABBAGE WHITEFLY

The cabbage whitefly is most likely to be found on brassicas where it pierces the plant and sucks up the juices. It then gets rid of any excess sugars taken up with the sap, leaving a sticky mess and ruining crops. Eggs are laid on the undersides of leaves and the first stage nymphs are quite active. However, the legs and antennae degenerate after the first moult and the cabbage whitefly is motionless in the next two stages. It continues to feed during this period and eventually reaches the pupal stage, when it stops feeding and develops adult appendages. The nymphs moult four times before they reach their adult form.

SIZE 2mm.
DESCRIPTION Wings covered with a white waxy coating. Resembles a minute moth.
HABITAT Gardens, fields. Found throughout the British Isles.
SIMILAR SPECIES Greenhouse whitefly, which is likely to attack tomatoes, cucumbers or house plants.

STAG BEETLE
(LUCANUS CERVUS)

The male stag beetle is Britain's largest beetle. If disturbed it will threaten with its antlers, but it is not actually capable of inflicting any damage as the muscles that move them are weak. It can be seen flying in the evening between June and August. After mating the female will go in search of some rotting wood on which to lay her eggs, usually choosing oaks or fruit trees. She dies shortly afterwards. The larvae spend up to three years feeding on the wood and then form pupae within it. They pass the winter there and hatch in the following spring. The adults live on fat reserves developed during the larval stage and have very short lives.

SIZE Male 40mm (excluding horns), female 30mm (no horns).

DESCRIPTION Black head and body, brown thorax. Violet colouring to wing cases. Antler-shaped extended jaws.

HABITAT Tree trunks and stumps, fences. Only present in south and central England.

SIMILAR SPECIES Lesser stag beetle, which is more common but smaller.

VIOLET GROUND BEETLE (CARABUS VIOLACEUS)

SIZE 20–30mm.
DESCRIPTION Black body and wing tips edged with purple, violet or red-blue. Feet have five joints.
HABITAT Woods, gardens, waste ground.
SIMILAR SPECIES Carabus granulatus, which has reddish edges to wing tips and rows of large granules on wing cases. Is found in damp places.

The violet ground beetle is one of the largest species of ground beetles, resting by day in cool, dark places and hunting at night. It is a carnivore, seeking out slugs and other plant-feeding insects, crushing its victims in its strong jaws. It cannot fly but has powerful legs, giving it the ability to move over ground very quickly. Its eggs are laid on plants in the spring and the resulting larvae live among plant litter and also eat soft-bodied insects. These larvae are fully formed at ten months and then develop into pupae inside rotten wood or in the soil. The adult emerges from the pupa in autumn but does not become active until the following spring, then having a life expectancy of about nine months.

LADYBIRD
(COCCINELLA)

The seven-spot ladybird is just one of the many different types of ladybird which all vary in their colour and markings. Its familiar bright colours are there to warn predators that it is unpleasant and dangerous to eat. It actually contains poisonous alkaloids and, to aid protection further, can also produce yellow blood from certain points on its body. The adult ladybird hibernates during the winter. Both ladybirds and their larvae prey on aphids. Eggs are laid on plants covered with aphids and the blue larvae consume hundreds of them during their three-week developmental period. They then form pupae and attach themselves to leaves or stems.

SIZE 6mm.

DESCRIPTION Small red beetle with seven black spots – three on each wing and a fused pair. Wing cases can be orange instead of the normal red.

HABITAT Widespread throughout the British Isles.

SIMILAR SPECIES There are over forty species of ladybirds with the number of spots varying from none to twenty-four. They all eat mildew or aphids.

BLACK VINE WEEVIL (OTIORHYNCHUS SULCATUS)

Size 8–12mm.
Description Brown-black in colour. Jointed antennae and broad snout.
Habitat Greenhouses, conservatories, parks and gardens.
Similar species Nut weevil, which is smaller with brown colouring. Lays its eggs in nuts.

Weevils have longer snouts (rostrums) than beetles and these all vary in shape. The black vine weevil larvae cause a great deal of harm to plants by feeding on tubers and roots to such an extent that the plant no longer has the ability to draw up water and dies. Once it reaches its adult form, the black vine weevil can also be a pest if it gets into greenhouses or onto pot plants. It eats the edges of leaves, marking them with semi-circular bites. The fact that it feeds at night makes it difficult to see and if disturbed it will deliberately fall off the plant and play dead. The poison used to control them was withdrawn in 1990 and this has led to an increase in their population. There are over five hundred different species of weevil.

(PHILAENUS SPUMARIUS) COMMON FROGHOPPER

The common froghopper is active between May and September. In the autumn the female lays between fifty and one hundred eggs in the crevices of dead stems. These hatch the following spring and the nymphs climb onto plants such as sorrel or hawthorn to suck up the sap. Air then bubbles up in a secretion from the abdomen and a frothy coat is developed, protecting the nymph from potential predators. It remains like this until fully formed when it is able to leap away.

SIZE 6–8mm.

DESCRIPTION Light brown with dark patches or dark brown with light patches.

HABITAT Hedgerows, gardens; any well-vegetated places with shrubs and herbaceous plants.

SIMILAR SPECIES Black and red froghopper, which has distinctive red and black markings.

COMMA BUTTERFLY (POLYGONIA C-ALBUM)

SIZE Wingspan 4.4–4.8cm.

DESCRIPTION Bright orange upper sides with dark brown spots and edges. Very jagged margins. Undersides are a mottled brown with a white 'comma' on the hind wing.

HABITAT Orchards, hedgerows, woodland and gardens. Is rare in Scotland.

The comma butterfly enjoys basking in the sunshine and tends to live a solitary life, drinking nectar from a variety of flowers such as asters, thistles or buddleias. It hibernates in hedges or hollow trees where it resembles a dead leaf. There are two broods each year, with butterflies that hibernated in winter mating in March or April and producing the first generation in July. These then mate to produce the next generation in September or October, which are darker in colouring. Eggs are laid on the upper side of hop leaves, nettles or currant bushes. The caterpillars resemble bird droppings, providing them with camouflage. The dark brown chrysalis hangs from its food plant.

PAINTED LADY BUTTERFLY

(CYNTHIA CARDUI)

In May and June the painted lady butterfly migrates to Britain from North Africa and south-west Europe, travelling over eight hundred miles. The numbers making the journey each year vary depending on weather conditions in the country of origin. Eggs are laid on the upper sides of leaves of plants such as nettles and thistles. The caterpillar has a yellow stripe down each side and yellow or black spines. The resulting chrysalis hatches after being suspended from the food plant for two weeks. Sometimes a second generation will be produced in September or October, but these butterflies cannot tolerate cold weather in any form and soon die.

SIZE Wingspan 5.4–5.8cm.
DESCRIPTION The upper sides are a pale orange and have black markings - there is a broad triangle at the wing tip with white spots within it. The hindwing underside is a mottled brown-grey with a row of blue eye spots.
HABITAT Gardens and anywhere containing flowers.

PEACOCK BUTTERFLY (INACHIS IO)

SIZE Wingspan 5.4–5.8cm.
DESCRIPTION Upper sides are a rich chestnut with four large, lilac-blue eye spots. Margins are sooty coloured. Undersides are mainly black.
HABITAT Gardens, parks and anywhere with flowers rich in nectar.

The peacock butterfly seeks out plants such as buddleias, thistles and ragwort and also feeds on rotten fruit in orchards. Its eye spots have evolved to deter predators such as birds and lizards and it patrols its territory, rubbing its wings together to ward off any enemies. In the winter it hibernates indoors, using ceilings or corners in outbuildings. It flies in early to mid-spring and from midsummer to early autumn. It lays its eggs on the undersides of nettles and the caterpillars feed on the plants. They are black and hairy with tiny white spots. The chrysalis is formed in places such as tree trunks and nettle stems and is suspended by its silk pad.

SMALL TORTOISESHELL BUTTERFLY

(AGLAIS URTICAE)

The small tortoiseshell is probably the most common butterfly in Britain. Its favourite plants include buddleia, candytuft, common thistle and aubretia. It is able to survive through the winter, hibernating in its adult form, and often hides in corners in houses, also using dry outbuildings. It re-emerges in March and eggs are laid on the underside of stinging-nettle leaves. The caterpillars are black and yellow and live in colonies until the end of their development, giving some protection from predators. The chrysalis forms under eaves and windowsills and is often vulnerable to blue tits. The first generation emerge as butterflies in June or July and these produce a second generation, appearing in August or September and then hibernating.

SIZE Wingspan 4.5–5cm.
DESCRIPTION Upper sides are orange and have a black border with blue spots within it. There are three rectangular patches at the front. The hindwing has a black base and the undersides are brown-black.
HABITAT Gardens, parks and any places with flowery plants.
SIMILAR SPECIES Large tortoiseshell, which is larger and less brightly coloured. It is very rare in the British Isles.

RED ADMIRAL BUTTERFLY
(VANESSA ATALANTA)

The red admiral is very territorial and regularly patrols small stretches of hedgerows or lanes, sending away other butterflies. It can be seen flying between May and October and may hibernate through the winter, although few survive. In gardens it feeds on plants such as buddleia and Michaelmas daisies; in the wild it sucks the nectar from clover, scabious and teasel. It also feasts on windfalls and drinks from puddles. The majority of red admirals in Britain are migrants from southern Europe. They arrive in May and lay their green, ribbed eggs on the upper sides of nettle leaves. When the caterpillar emerges, it is dark with white speckles and yellow spots along the sides. It makes a tent of leaves over itself, using silk threads to pull them together. It stays there feeding before moving onto the next leaf where it makes a larger home. The chrysalis is formed during summer or autumn and is often suspended from the same plant. About seventeen days later the butterfly emerges.

SIZE Wingspan 5.6–6.2cm.
DESCRIPTION Velvety upper sides with vivid red and white markings. Undersides are dark brown and mottled with black and blue.
HABITAT Parks, gardens and any other places with an abundance of flowers.

WALL BROWN BUTTERFLY (LASIOMMATA MEGERA)

SIZE Wingspan 3.6–5cm.
DESCRIPTION Upper sides are deep orange in colour with brown streaks and borders. There is an eye spot on each forewing with an orange ring round it. Three or four smaller eye spots are present on each hind wing. Undersides of hindwings are mottled brown and have three eye spots on each one.
HABITAT Gardens, rough grassland, heathland, open woodland.

The wall brown butterfly derives its name from its tendency to bask on walls in the sunshine. The male is very territorial, constantly looking out for and driving away insects that try to intrude. Two broods are produced each year; the first in May or June and the second in July or August. Very occasionally, three may be produced. Eggs are laid on the undersides of grasses and the caterpillars that emerge are blue-green and hairy, mainly feeding at night. In early summer they develop in a month but later in the year they will hibernate through the winter until April. The chrysalis hangs from a stem and the butterfly emerges after two weeks, having a lifespan of about three weeks.

(ANAX IMPERATOR) EMPEROR DRAGONFLY

The emperor is Britain's largest dragonfly and is only found in south-east England. The male is strongly territorial and will fight off other males, sometimes leading to fights in mid-air. This dragonfly can be seen flying from June to September. It feeds on other insects, catching them on the wing by extending its forelegs forward in a trap. It constantly flies up and down stretches of water looking for food, reaching speeds of eighteen miles an hour. After catching a victim, it settles on a convenient perch to eat. The male and female mate while flying and the male supports her as she lays her eggs on plants beneath the surface of the water; he then helps her up again. The nymphs hatch and spend two years in the water eating food, including tadpoles, using a lower lip extension that has two hooks to catch prey. Once fully grown, they leave the water in spring and cling to a plant as the adult emerges; they live for a further month.

SIZE Body length 70–85mm. Wingspan 100–110mm.

DESCRIPTION Abdomen of male is sky blue and the female's is green-blue. Both have a dark stripe running the length of the back.

HABITAT Ponds, canals and lakes.

SIMILAR SPECIES Common aeshna, which is smaller. Males have blue spots on abdomen and females have green spots.

COMMON EARWIG (FORFICULA AURICULARIA)

SIZE Body length 10–15mm (not including pincers).

DESCRIPTION Body is a shiny chestnut brown and has pincers at tail end which are curved on the male and straight on the female. Has short, leathery forewings which cover the hindwings which are folded.

HABITAT Trees, gardens, houses. Present throughout the British Isles.

SIMILAR SPECIES Lesser earwig, which is smaller (5mm) and a paler colour.

The earwig is predominantly vegetarian and often chooses to feed on the petals of plants such as dahlias and chrysanthemums. It can fly well, but has to fold its wings carefully back into the wing cases by using its pincers after doing so. The male and female do not mate until autumn, after which they hibernate for the winter in a cavity in the soil situated at the base of a plant. The female lays her eggs in the spring in an underground cell. She looks after them carefully, guarding them and licking them to keep any bacteria away. They hatch after three or four weeks and she brings the young earwigs food for the first few days.

(SYRPHUS RIBESLI) HOVERFLY

The hoverfly physically resembles a wasp although its flight and sound are completely different. The male occasionally makes a droning noise and has the ability to hover motionless in the air. This mimicry is a form of defence against predators who mistake it for a wasp and are consequently wary. The hoverfly can be seen from April to November and feeds on nectar and pollen, so will populate any area rich in flowers. It often, lazily, follows the bumble bee, which bites holes in the pollen tube. In this way the hoverfly can access pollen in flowers where it is normally too awkward for it to reach, such as the common comfrey. The larvae develop in ants' nests and are sometimes known as red-tailed maggots.

SIZE Body length 12mm.
DESCRIPTION Fly with black and yellow striped markings.
HABITAT Gardens, parks, and woodland. Widespread throughout the British Isles.
SIMILAR SPECIES There are about 250 different species of hoverfly, all with slightly different markings.

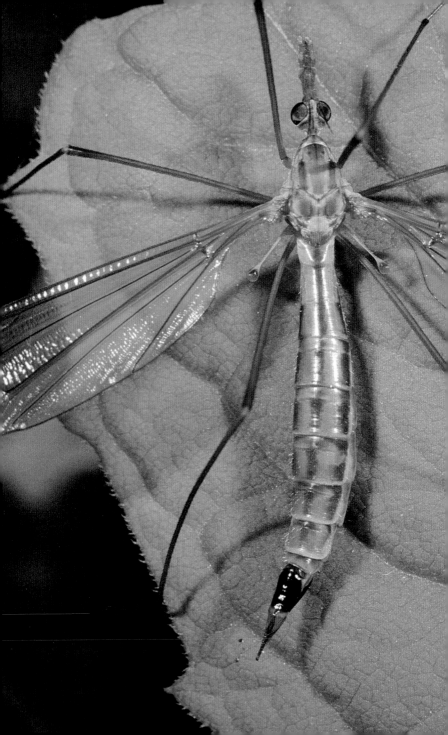

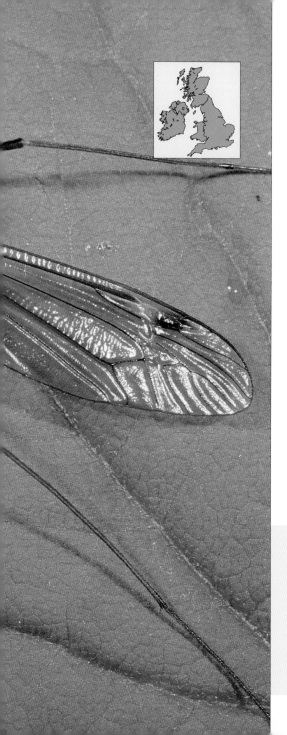

CRANEFLY
(TIPULA PALUDOSA)

The cranefly or daddy-long-legs, as it is more commonly known, can be seen in the late summer or autumn and has a very familiar dancing flight pattern. It is drawn to artificial light and will often come indoors; in the daytime it is more likely to be seen on lawns or pastures. The eggs hatch after fourteen days and the larvae are more commonly known as leatherjackets. They survive in the soil during the winter and are a serious pest to farmers and gardeners, eating roots and stems and potentially killing patches of lawn. Their predators include rooks and other birds.

SIZE Body length 30mm.
DESCRIPTION Long, thin body with transparent, shiny wings and long, very fragile legs.
HABITAT Grassland, gardens, roadsides, farmland.
SIMILAR SPECIES Tipula maxima, which is larger and usually seen near water.

NON-BITING MIDGE (CHIRONOMUS PLUMOSUS)

SIZE 8mm.

DESCRIPTION Slender body and transparent wings.

HABITAT Woods, ponds, ditches, gardens and farmyards.

SIMILAR SPECIES There are nearly four hundred different species of non-biting midge.

Midges are widespread throughout the British Isles and the males often congregate in large swarms near water. The females are attracted to these swarms and mating takes place on the wing. The larvae are bright red and live in the bottom of ditches, wriggling to make a figure-of-eight shape. They are often known as bloodworms.

(THEOBALDIA ANNULATA) RINGED MOSQUITO

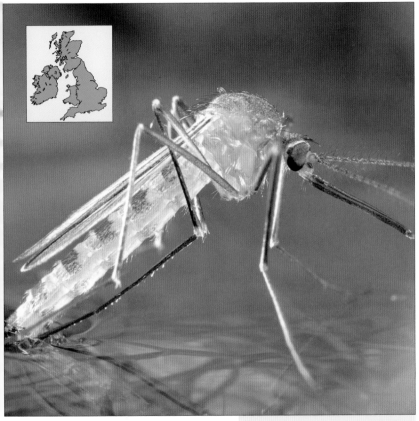

The ringed mosquito mates in the autumn and the male dies soon afterwards, while the female hibernates for the winter. She finds a suitable building and occasionally emerges if the weather is warm enough. In spring she searches for stagnant water on which to lay her eggs. The larvae live just below the surface of the water and feed on floating food matter. Mosquitoes generally feast on blood from animals or birds, but the female ringed mosquito is very much attracted to human blood.

Size Body length 9mm.
Description Has white rings on the legs and wings with dark spots.
Habitat Winter spent hibernating in buildings. Summer spent near water where it breeds.
Similar species Common gnat, which is slightly smaller and does not bite humans.

HOUSE CRICKET
(ACHETA DOMESTICUS)

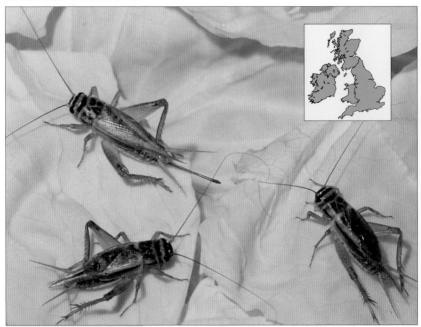

SIZE Body length 15–20mm.

DESCRIPTION Body colouring yellow-brown or grey with darker markings. Has large wings and long antennae. It appears to have an extra tail which is actually the tightly rolled hindwings, which project beyond the two cerci.

HABITAT Heated buildings, especially kitchens or small bakeries, rubbish dumps.

SIMILAR SPECIES Dark bush cricket, which is more brown-black in colour and has much longer antennae.

The house cricket is actually a native of north Africa and cannot tolerate cold conditions, hence its need to live indoors or on rubbish heaps where the fermentation process provides heat. It lives in warm crevices by day and at night emerges to feed. Mating also takes place at night when the male attracts the female, emitting a long chirp by rubbing the right forewing against a toothed rib in the left forewing. The female has the potential to lay up to one thousand eggs during her lifespan of six to nine months, and deposits them in crevices in walls or wood. Depending on the surrounding temperature, the eggs take from one week up to four months to hatch into the larvae (nymphs). These then develop into adults after one to eight months; again, depending on warmth. The young nymphs are almost identical to the adults but have no wings.

COMMON FIELD GRASSHOPPER
(CHORTHIPPUS BRUNNEUS)

During the day the common field grasshopper needs to warm up in the sunshine, basking until it can become active. The male chirps during the months between June and November and achieves this sound by rubbing pegs on his hind legs against larger veins on his forewing, a process called stridulating. This is to attract the female who, after mating, will lay up to fourteen eggs in a foaming secretion which protects them. The eggs stay like this through the winter and hatch the following spring. In their larval form as nymphs they are very similar to the adults, but without wings. The nymphs undergo about four moults and three months later emerge as adults.

SIZE Body length 18–24mm.
DESCRIPTION Body colouring is brown and green although shades will vary. Short antennae and very long forewings.
HABITAT Open woodland, fields and grassy places.
SIMILAR SPECIES Meadow grasshopper, which has much shorter forewings and cannot fly.

GREEN LACEWING (CHRYSOPA PERLA)

SIZE Body length 15mm.
DESCRIPTION Wings are transparent with obvious veins. Black spot between the antennae.
HABITAT Mainly found in southern England and absent from Scotland. Found in woods, hedgerows and gardens.
SIMILAR SPECIES Chrysopa carnea, which turns bright pink in winter.

The green lacewing is seen between April and October. Its main diet consists of aphids, including greenfly, so it is a positive bonus in the garden. If threatened by predators, the lacewing can secrete a foul-smelling liquid from its stink glands. The female lays her eggs by secreting sticky threads from her abdomen, making very thin stalks; she then puts an individual egg on the end of each one. These can be seen on the underside of plants such as roses and Michaelmas daisies. The first larvae to hatch eat any unhatched eggs and then seek out aphids. They use their long mouthparts to suck out the contents of the aphid's body by piercing a hole in it first. An individual larva then uses the aphid's skin as a home in which it can hide. It spins a silk cocoon before developing into an adult. Usually two or three generations a year are produced.

(SATURNIA PAVONIA) EMPEROR MOTH

The male emperor moth flies during the day whereas the female is active at night. The male searches for a female and can pick up her scent from two kilometres away, as the newly hatched female produces a very large amount of pheromones. This often results in many males being attracted to one female. After mating she lays a ring of eggs on the stems of plants such as bramble or heather. The caterpillars that hatch in May or June are black and hairy and usually feed together. As they grow, their colour changes to green and they develop yellow, hairy warts. In July or August a cocoon is spun within the foliage of the food plant and the moth finally emerges the following May. Although it is a member of the silk moth family, emperor moth cocoons are unsuitable for the production of silk.

SIZE Forewing of female is 40mm. Forewing of male is 21mm.

DESCRIPTION Large prominent eye spots on each wing. Females have grey wings and males have grey-brown forewings with tawny-coloured hindwings. Male has feathery antennae.

HABITAT Widespread throughout the British Isles. Found in heaths, grassland, moors and open woodland.

GARDEN TIGER MOTH
(ARCTIA CAJA)

This colourful and distinctive moth only flies at night in July and August, so is rarely seen. The garden tiger moth uses its bright colours and distinctive smell to warn off potential predators. It can also produce a grating noise by rubbing its wings together, and can secrete yellow blood from the thorax. The eggs are laid on the underside of plant leaves such as dandelion, dock and nettle. The extremely hairy caterpillars that emerge are frequently referred to as 'woolly bears' and can be found in May or June. Hibernation takes place during the winter, and the caterpillar emerges the following spring to make a cocoon in leaf litter. This is made from silk mixed with its own hair and eventually changes into a glossy chrysalis.

SIZE Forewing 35–39mm.
DESCRIPTION Brown and white forewings and red hindwings with black spots. Colours and markings are variable so no two are exactly alike. Females are larger.
HABITAT Found in gardens throughout the British Isles.

ELEPHANT HAWKMOTH
(DEILEPHILA ELPENOR)

SIZE Forewing 31–36mm.

DESCRIPTION Forewings are deep pink with olive-green markings. Hindwings are bright pink and have black undersides. White antennae and legs. Females are larger.

HABITAT Widespread throughout the British Isles except for north Scotland. Found in gardens, parks and open woodland.

SIMILAR SPECIES Small elephant hawkmoth, which is slightly smaller and lacks the black colouring on the hindwings. The green colours are more yellowish in this species.

The elephant hawkmoth is able to see in colour, even when it is so dark that humans cannot see at all. It flies in June and spends the evenings feeding on flowers such as honeysuckle, petunias and valerian. The eggs are laid individually on willowherbs, and the caterpillars are green or dark brown with two sets of false eye spots behind the snout. The caterpillar's tapered front, resembling an elephant's trunk, has given this moth its unusual name. When alarmed the caterpillar retracts the 'trunk' and swells the eye spots, which makes the front of the body look even more like an elephant's head. It is harmless and can be seen in July and August, usually feeding at night and basking on plants during the day. It forms a chrysalis just below the surface of the soil where it spends the winter.

LARGE WHITE PLUME MOTH
(PTEROPHORUS PENTADACTYLA)

The large white plume moth is completely white in colour and spends the day resting on long grasses. It is very attracted to light. The wings are divided into lobes which gives them a feathered appearance. The caterpillar feeds on bindweed and hibernates until the spring when it feeds on young flowers and leaves.

SIZE Forewing 12–15mm.

DESCRIPTION The forewings are divided into two lobes and the hindwings are divided into three. They are feathery in appearance and join together to rest at right angles to the body. They have long thin legs and bodies and the hindlegs have side spurs. These features give the moth the appearance of the letter 'T'.

HABITAT Found in gardens, waste ground and hedgerows.

SIMILAR SPECIES Many-plumed moth, which is brown and has wings with ten lobes.

MAGPIE MOTH (ABRAXAS GROSSULARIATA)

SIZE Wingspan 35–40mm.

DESCRIPTION Upper sides have a white background with varied black and yellow patterns. Undersides are less marked. Yellow body. Females are larger.

HABITAT Gardens, hedgerows, wasteland and fruit farms.

SIMILAR SPECIES Clouded magpie, which has similar, more blurred markings and feeds on elm and beech leaves.

The magpie moth has gained its name from its markings, which are similar to those of the magpie. It flies during the day in July and August and lays its eggs on the undersides of the leaves of hawthorn, currant, gooseberry and blackthorn bushes. These hatch after ten days and the caterpillars are white with black spots on their backs and a yellow stripe down each side. After hibernating through the winter they emerge in May or June to feed from the plant on which they were laid, and are quite capable of stripping the leaves from fruit bushes, thus destroying them. They form a distinctive chrysalis, striped with black and yellow, which can be found on currant leaves in early July, when they finally hatch.

(CERURA VINULA) PUSSMOTH

The pussmoth's name comes from a covering of grey, cat-like hair on its abdomen. It flies from April to July and lays reddish-coloured eggs on the upper side of willow, sallow or poplar leaves. The caterpillar that hatches is green with a purple 'saddle' surrounded by a white border. It has white and purple spots on the body, a large, red face with unique eye markings and two tails that it waves when threatened. It also has the ability to squirt an intruder with formic acid. It can be seen in late summer before weaving its cocoon in August. Bark is mixed with silk to give good camouflage, the chrysalis then spending the winter in the cocoon on the side of a tree. Using a cutting device on the head it finally breaks through the cocoon, enabling it to emerge as a fully-formed adult.

SIZE Wingspan 60–80mm.
DESCRIPTION Upper sides are pale grey and white with dark swirls and wavy lines. Abdomen is grey and furry.
HABITAT Found in open woodland and hedgerows close to willows, poplars and sallows.

RED UNDERWING MOTH
(CATOCALA NUPTA)

SIZE Wingspan 60mm.

DESCRIPTION Forewings are grey-brown with a darker marbling pattern. Hindwings have black and scarlet bands with a white margin.

HABITAT Mainly found in south and central England. Absent from Ireland and Scotland. Found in woodland, parks, gardens and marshes.

SIMILAR SPECIES Dark crimson underwing, which has more crimson-purple and black hindwings.

This moth uses its colours to escape predators. When flying, the red underwing moth reveals the crimson on its hindwings but then settles on a tree where it closes its wings. The forewings then blend in with the tree's bark. It is one of Britain's largest moths, rests by day and flies at night, and is strongly attracted to rotten fruit. It can be seen flying in August and September. After mating, eggs are laid in crevices in the bark of willows or poplar trees. The caterpillars are grey and feed at night on the leaves of the breeding tree between April and July. The chrysalis forms its cocoon in the bark or amongst leaf litter where it stays for the winter.

HAWTHORN SHIELD BUG

(ACANTHOSOMA HAEMORRHOIDALE)

Britain has over 1500 species of true bugs most of which feed on plants and thrive in the drier southern and eastern areas of the country. The hawthorn shieldbug, a terrestrial bug named because of its shield-like shape and its preference for the fruit of the hawthorn tree, is common throughout England and Wales but is not found in Scotland. It is present during the autumn and then hibernates through the winter in leaf litter. In the spring the female lays her eggs on the undersides of hawthorn leaves and the larvae hatch nine days later.

SIZE 15mm.

DESCRIPTION Triangular, flat beetle with a bright green shield with red markings.

HABITAT Parks, hedges and shrubs where hawthorn trees are present.

SIMILAR SPECIES Green shield bug, which is found on hazel trees and beans. It is a bronze colour in the autumn, turning to bright green in the spring.

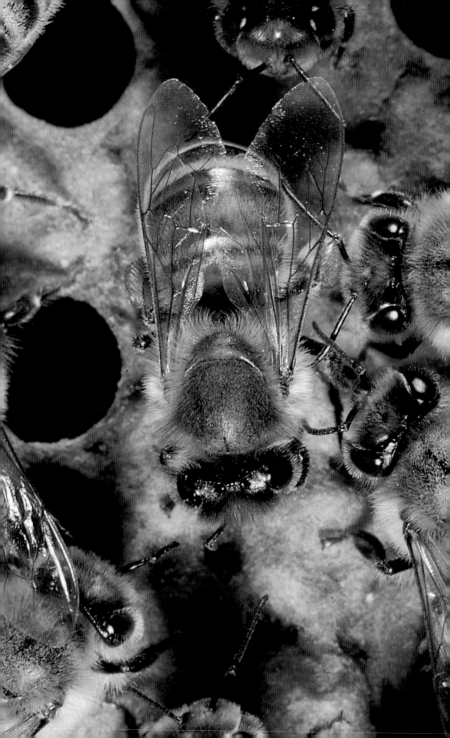

HONEY BEE (APIS MELLIFERA)

Honey bee colonies are a fascinating mix of bees taking on many different roles. There are the bees that make and provide honey and the workers (the sterile females) are those responsible for looking after the hive. They carry out various duties including cleaning and guarding it, feeding the larvae and gathering the food necessary for all the bees' constant demands. The honey, produced along with pollen, is used to feed both the larvae and a reduced population of bees through the winter. Forty pounds of honey can be produced by a colony of 50,000 bees in a single summer. Workers secrete wax from glands to make the honeycomb, which is then secured with resin collected from trees. Nectar is gathered from plants and regurgitated by the workers once back at the honeycomb. This turns to honey. Pollen is also collected and carried back to the hive. The queen bee makes her first flight when she hears the noise of a group of males (drones). She mates in mid-air and the drone immediately dies. She then secretes a substance over her body, which is licked by the workers. This makes them all work as a colony and stops new queens being reared. She then lays one egg in each cell during April, finally laying up to two thousand. The eggs change to larvae and then to pupae. Eventually the colony grows to a size where some workers cannot reach the queen's secretion. New queens then emerge who leave with swarms of workers between May and July to form new colonies.

SIZE 12mm.

DESCRIPTION Hairy body with brown and black stripes. Transparent wings.

HABITAT Widespread throughout the British Isles in a variety of habitats.

SIMILAR SPECIES Flower bee, which has a yellow face and long hairs on the middle legs.

BUFF-TAILED BUMBLE BEE (BOMBUS TERRESTRIS)

SIZE 25mm.
DESCRIPTION Black body with two orange bands and a white tail.
HABITAT Widespread throughout the British Isles except in northern Scotland.
SIMILAR SPECIES Common carder bee, which is smaller and without the coloured bands on the body.

The buff-tailed bee is the largest and most common species of British bumble bee, forming colonies consisting of workers, drones and a queen, although in much smaller numbers than the honey bee. An average colony will only consist of about one hundred and fifty bees who cannot survive the winter as they do not produce sufficient honey for food. The drones and workers die in the autumn while the young queens hibernate and emerge in the spring, setting up new colonies in hedgerows or in a nest previously belonging to a mouse. Buff-tailed bumble bees reach nectar and pollen by weighing down the flower to open the petals or biting a hole in the flower tube so they so they can reach inside with their tongues.

(PHALAGNIUM OPILIO) HARVESTMAN

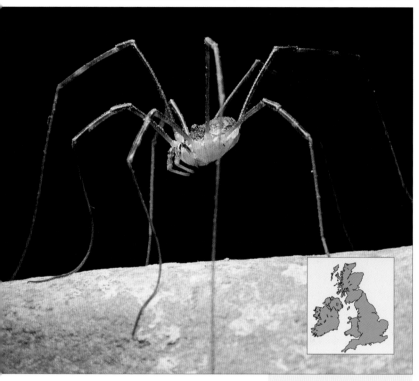

Although the harvestman looks like a spider, it is only vaguely related. The main differences are that the spider has a divided body and is able to spin a web; the harvestman has a one-piece body and no silk glands to build such a structure. It gained its name from a habit of walking over harvested fields in late summer. Its main diet consists of small insects and plant material; it feeds at night and basks during the day. It lives from four to nine months and can defend itself from predators by either secreting an unpleasant fluid or shedding a leg, which does not grow again.

SIZE 3–6mm.
DESCRIPTION Eight long legs with the second pair being the longest. Small roundish body. Female has a long ovipositor to lay her eggs in the ground.
HABITAT Verges, gardens, grassland.
SIMILAR SPECIES Other harvestmen such as homalenotus quadridentatus, which has four spikes on its body and back.

GARDEN SPIDER (ARANEA DIADEMATUS)

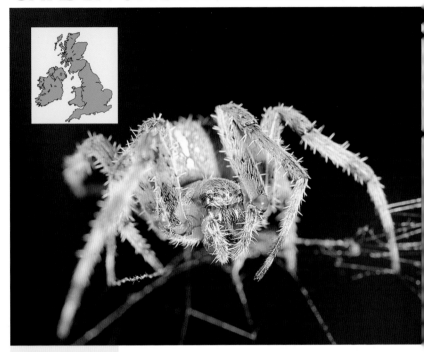

SIZE 12–15mm.
DESCRIPTION Brown body which is occasionally black or orange, and pale cross-shaped marks on back.
HABITAT Hedgerows, gardens.
SIMILAR SPECIES Araneus quadratus, which does not have the cross-shaped marking.

The garden spider is one of the forty species of orb-web spiders, which spin traditionally shaped webs. The webs are made by spinning a non-sticky silk that the female produces from a gland at the tip of her body. They are attached to plants and consist of an outer spiral designed to catch unsuspecting prey and an inner part where the female sits and waits. She then eats the outer spiral, producing gummed silk to replace it. Once caught, the prey is paralysed by a bite from the spider, which injects it with enzymes to turn it into a liquefied form for eating. If the web is damaged beyond the point where it can trap prey, the contents are eaten and the female spider starts again. The male will feed from the female's web but does not actually share it. She lays eggs in the autumn in one large mass, containing as many as eight hundred, which is protected by yellow coloured silk. She dies soon afterwards. The eggs hatch in the spring but remain in a cluster at first.

(TEGENARIA DOMESTICA) HOUSE SPIDER

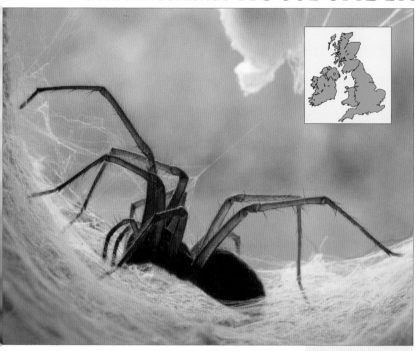

The house spider is able to spin a very large web, which is as much as thirty centimetres in diameter, and then waits in the corner until it feels the vibration from any small insect that has been caught. The web is not actually sticky; the insect's feet become entangled in it. The prey is then poisoned by the spider and eaten. This spider is actually very useful in houses, as it keeps fly populations down. It is particularly common in late summer when it is fully developed and at its fastest. It has a pair of palps at the front, the equivalent of antennae. These help them source food, and just in front of them are the pincers which assist with tearing the food. The house spider cannot take in solids so either sucks juices or liquefies its food before eating. The male uses the palps to fertilise the female when he inserts the sperm.

SIZE 10–18mm.
DESCRIPTION Brown body with black mottling. Long and slightly hairy legs.
HABITAT Sheds, garages, houses.
SIMILAR SPECIES Cardinal spider, which is larger.

WATER BOATMAN (NOTONECTA GLAUCA)

SIZE 14mm.

DESCRIPTION Brown body with dark markings. Hind legs are long and fringed. Very large eyes.

HABITAT Common throughout British Isles in still water such as ponds and lakes.

SIMILAR SPECIES Lesser water boatman, which is smaller and swims the right side up.

The water boatman actually swims upside down with the surface tension suspending it just under water. It uses its hairy, long, hind legs as paddles so it looks as if it is rowing. The swimming position is actually controlled by light rather than gravity; experiments in tanks have shown that a light shone from underneath will make it swim the right way up. It feeds on insects, small fish and tadpoles and can be quite aggressive in the chase. Both the eyes and vibrations in the water are used to look for food. The water boatman is a bubble-breather, so it surfaces regularly for air. It is present throughout the year and flies well.

(ONISCUS ASELLUS) WOODLOUSE

The woodlouse is actually related to crabs and lobsters but has adapted to life on land and lives under logs and stones. Its body has never actually developed to be fully waterproof so it needs damp conditions to avoid drying out in sunshine. It tends to be active at night when it is both safer from drying conditions and potential predators. It feeds on fungi, plant leaves and decaying plant matter and eats its own droppings, which still contain nourishment. The female lays her eggs into a pouch under her body where they hatch after three to five weeks. They are fully mature after two years.

SIZE 14mm.
DESCRIPTION Body is dark grey and divided into visible segments with pale sides to the shell.
HABITAT Hedgerows, woodland, gardens, parks.
SIMILAR SPECIES Pill woodlouse, which is grey and has no markings. It has the ability to roll itself into a ball if disturbed.

COMMON CENTIPEDE
(LITHOBIUS FORFICATUS)

During the day the centipede rests in hidden places such as leaf litter or under stones to maintain its moisture level. It does not have a waterproof layer and therefore only emerges at night when there is no danger of drying out. As it lives in virtual darkness it does not have eyes and instead uses touch and vibration to catch its prey. It feeds on creatures such as woodlice and insects; once it has caught something, it injects it with poison contained within its front legs. The young are produced from eggs and they have a hard outer skeleton which is regularly shed to allow for growth.

SIZE Body length up to 30mm.

DESCRIPTION Brown body that is divided into shiny-looking segments. They begin with seven pairs of legs but, after several moults, adults have fifteen pairs.

HABITAT Widespread throughout the British Isles.

SIMILAR SPECIES Cryptops hortensis, which is shorter, paler and usually found in dead wood.

FRESHWATER CRAYFISH (ASTACUS PALLIPES)

SIZE Body length up to 10cm.
DESCRIPTION A crustacean with a brown or olive-green body that looks similar to a lobster.
HABITAT Streams and rivers.
SIMILAR SPECIES Signal crayfish, which originates from North America.

The freshwater crayfish is also known as the white-clawed crayfish and is a true British native. It lives in fast-moving water that is thoroughly oxygenated, preferably having run across limestone or chalk. Its main diet consists of snails, insect larvae or tadpoles and it feeds at night, hiding under stones during the day. The eggs, when fertilised, are carried by the female and once the young hatch, they use their pincers to cling onto their mother. Freshwater crayfish numbers are now under threat due to the introduction of the signal crayfish from America, but also because of pollution and the disturbance of preferred streams and rivers.

(GAMMARUS PULEX) FRESHWATER SHRIMP

The freshwater shrimp is more common in chalk and limestone areas and feeds on detritus usually found under stones, filtering the water for food. It moves by swimming on its side, having a body that is laterally flattened. Although it jerks along, it is actually quite a fast swimmer. Three pairs of legs at the front move constantly which provides the gills with a good supply of oxygen-rich water. Predators of the freshwater shrimp include birds, fish and some insect larvae. The male often carries the female in the breeding season when she keeps her eggs inside a brood pouch in the body; the young are then hatched live.

SIZE Body length up to 3cm.

DESCRIPTION Has a segmented, curved body that is almost translucent and light brown in colour.

HABITAT Streams, rivers and ponds that are clean – cannot tolerate polluted water or low levels of oxygen.

SIMILAR SPECIES Fairy shrimp, which is totally transparent and swims upside down. Very rare.

COMMON SNAIL
(HELIX ASPERSA)

The common snail is also referred to as the garden snail, and is found in virtually all gardens where there is alkaline soil. It rests during the day to avoid warmth and predators, and is active at night, feeding on various plants. Snails have both male and female reproductive organs and the sperm is shot out during mating to reach another snail which rests alongside. Eggs are laid in small holes in the soil; up to forty are produced at a time. The young hatch after a month and are miniature versions of the adults. In very dry conditions snails are able to seal their shells with mucus to avoid the loss of the moisture they need.

SIZE Body length 3cm.
DESCRIPTION Shells are a spiral shape with tortoiseshell markings that vary.
HABITAT Hedges, gardens, woodland.
SIMILAR SPECIES Roman snail, which has a cream-coloured shell.

GREAT POND SNAIL (LIMNAEA STAGNALIS)

Size Up to 50mm in height.
Description Shell is brown and conical in shape.
Habitat Canals, ponds and lakes. Mainly found in England and Wales.
Similar species Wandering snail.

The great pond snail's main diet consists of algae and other animal or vegetable matter and it feeds by rasping along surfaces with its tongue; those kept in tanks often leave a feeding trail. It prefers to live in stagnant water and can be seen gliding just under the surface film or moving amongst the plants. It lays its eggs on the underside of waterweed leaves, where they look like an oblong mass of gelatine. The young hatch after a month and immediately seek out the surface of the water to provide them with air. They are mature after two years.

COMMON EARTHWORM

(LUMBRICUS TERRESTRIS)

The earthworm is useful as it helps to keep the soil fertile. An acre of grassland can potentially contain up to three million earthworms, which live in tunnels as deep as 1.8 metres. The earthworm survives by feeding on soil, digesting any organic matter and bacteria within it and then excreting it in the top section, returning minerals to the upper parts of the soil. Its movement also introduces humus deeper into the earth. The earthworm has both male and female reproductive organs and mates at night on the soil's surface. Up to twenty eggs are laid inside the worm, wrapped in mucus, which then moves along the body to form a cocoon.

SIZE Body length up to 30cm long.

DESCRIPTION Soft pliable body with more than one hundred segments.

HABITAT Widespread throughout the British Isles, found in farms and gardens.

SIMILAR SPECIES Lumbricus rubellus, which is smaller and redder in colouring.

A FIELD GUIDE TO THE WILDLIFE OF THE BRITISH ISLES

FISH

THREE-SPINED STICKLEBACK
(GASTEROSTEUS ACULEATUS)

SIZE Length 4–10cm.

DESCRIPTION Silvery colour with green-blue back. In spring breeding periods males have bright red underparts and females become more yellow. There are three spines on its back before the dorsal fin. No scales, but bony plates on sides which can be along the whole length, at the front and tail or only at the front.

HABITAT Found in ponds, lakes, ditches and rivers. Additionally coastal waters and estuaries. Most common in the south and east of England.

FEEDING Insects, small worms, some crustaceans such as Daphnia.

SIMILAR SPECIES Ten-spined stickleback, which has eight to ten spines on its back and is darker in colour. In the breeding season the male has a black throat.

One of the most common freshwater fish, the stickleback is usually caught with a bent pin and a worm. At spawning time in May or June, it migrates upstream or moves to fresh water. The male then defends his chosen territory and makes a nest from detritus and vegetation, binding it all together with sticky threads that are secreted by the kidneys. He then attracts the female into the nest where she lays her eggs, which are immediately fertilised. Several females might lay in the same nest and the male guards the eggs for up to twenty-five days, fanning currents of water over them until they hatch. The young number sixty to a hundred and twenty; the male feeds them all for two weeks until they are independent.

MINNOW
(PHOXINUS PHOXINUS)

The minnow is generally caught on small worms or maggots or in a net laid on the bottom. It is then used as bait to attract larger fish such as trout. It is very common in Britain and lives in large shoals. The breeding season lasts from March to June when large numbers of these fish congregate in shallow water where there is sand and clean gravel. At spawning time the male grows white, horny tubercles on the head and his undersides become red. The female lays between one and six hundred eggs, which stick to the stones.

SIZE Length 8–10cm

DESCRIPTION Slim and small with a blunt-ended snout and small scales. Brown and green barred back and sides. No barbels. Cream undersides change to red in breeding season.

HABITAT Found in fast-flowing fresh water and vegetated zones.

FEEDING Small crustaceans, larvae and molluscs.

SIMILAR SPECIES Gudgeon, which is longer and lighter in colour with small barbels.

ATLANTIC SALMON
(SALMO SALAR)

The life cycle of a salmon is fascinating and begins with an egg laid in the gravel at the upper end of a river. After hatching, it stays there until it is a two-year-old smolt when it descends down the river to the sea. It swims across the North Atlantic to feed off Greenland. It remains there for any length of time between fourteen months and five years before returning to the river of its birth, often nearly to the exact spot. This precise journey covers thousands of miles before the salmon locates the river and majestically leaps and swims upsteam through waterfalls and rapids. It arrives at the spawning areas by October and then rests in nearby pools. In November or December the female moves up to the spawning area. She makes a nest by lifting stones with her tail to create a hole about fifteen centimetres deep. The male then joins her and they simultaneously shed eggs and milt. The milt contains millions of spermatozoa which fertilise the eggs, which the female then covers with stones. She makes up to a total of eight holes to deposit as many as 10,000 eggs. The eggs hatch after about a hundred days but the young fish stay in the stones for a month feeding on the egg yolks. After this period, they emerge from the stones and begin to feed on minute animals. Once the parents have entered the river to spawn they stop eating; after spawning they are extremely weak and survival rates are only between two and five per cent. Any that do survive return to the sea and may come back again two or three years later. It is believed that the salmon navigates across the ocean by salinity, water temperature and possibly also by magnetic fields. It then uses the trail of bile salts left by departing smolts to trace its birth river.

Size Up to 130cm.
Description Silvery body with dots on upper sides. Sharpish snout and small dark adipose fin. Belly is silvery-white with a pink tinge which changes to red at spawning time.
Habitat Rivers and sea.
Feeding Small fish, insects, shrimps in fresh water and shrimps, sprat and herrings at sea.
Similar species Trout, which is smaller and has fatter wrist at tail.

ROACH (RUTILUS RUTILUS)

Size Up to 50cm.
Description Silvery, deep-bodied fish
with a greenish tinge to their
backs. Reddish fins and red eyes.
The front edge of the dorsal fin is
immediately above the base of the
pelvic fin.
Habitat Very common fish living in
slow-flowing rivers and lakes.
Feeding Invertebrates such as
molluscs and plant material. Feeds
on the bottom.
Similar species Rudd, which has
golden tinge to back. Front edge
of dorsal fin is well behind base of
pelvic fin.

The roach lives in schools and can tolerate
poor-quality water. It can reach up to a
kilogram in weight and is very popular with
anglers, who use bread, maggots or worms
as bait. It spawns from April to June,
attaching its eggs to weeds with
5,000–200,000 eggs laid. The male matures
after two to three years and the female after
three to four. Maturity and growth rates
vary according to the quality of the water.

(TINCA TINCA) TENCH

The tench often appears lethargic but is a resilient fish with great strength. It requires very little oxygen and so is able to cope with stagnant or polluted water. It can also survive if the water source dries up, digging itself into the mud until the water begins to rise again. It can survive like this for some time and can even last up to three hours in a damp sack. The size of tench has begun to increase and it is thought that fertiliser from farms has got into the water, improving the quality of weeds and the invertebrates on which it feeds. It breeds from mid-May to July, congregating in groups and laying eggs amongst plants which then adhere to the weeds. One tench can lay up to 800,000 eggs. The fish are mature at the age of three or, occasionally, two years.

SIZE Up to 50cm.

DESCRIPTION Back and sides are blackish-brown or very dark green with a bronze lustre. Pink undersides and a barbel at each corner of the mouth. Very thick at tail. Reddish eyes.

HABITAT Likes sluggish water in muddy ponds and lakes.

FEEDING Feeds in the mud catching small fish, snails, mussels and insect larvae.

COMMON CARP
(CYPRINUS CARPIO)

The Romans originally introduced the common carp into Britain from Asia. It is often bred for fishing and for ornamental ponds. In the wild it has a more slender shape with even-sized scales but is normally bigger when cultivated, with a greater variety of scale patterns. To cope with colder water temperatures in winter, it digs into the silt and ceases feeding. Spawning takes place from May to July in shallows when the water temperature exceeds 18°C. The sticky eggs, numbering up to two million, are deposited on the leaves of water plants.

SIZE Length usually 25–60cm but can reach up to 100cm.

DESCRIPTION Deep-bodied fish with olive-gold body and even-sized scales. Long dorsal fin and prominent pectoral, pelvic and anal fins. Two barbels at each side of the mouth; thick, leathery lips.

HABITAT Still, warmer waters in the south that have muddy bottoms and plenty of vegetation.

FEEDING An omnivorous fish that feeds naturally on worms, larvae and snails but will also take bait such as meat, bread, potatoes and tinned pet food. It sucks up mud that passes through the alimentary canal while the food is digested and absorbed.

SIMILAR SPECIES Crucian carp, which has a more rounded top edge to the dorsal fin and is browner in colour. It has no barbels by the mouth.

COMMON BREAM (ABRAMIS BRAMA)

SIZE Length up to 30–40cm but can reach up to 75cm.

DESCRIPTION Deep, flattened body with a humped back. Grey-brown above and silvery below.

HABITAT Likes slow-flowing, brackish water, lakes and the lower reaches of rivers.

FEEDING Feeds on the bottom on invertebrates such as insect larvae, worms, pea-mussels and snails.

SIMILAR SPECIES Silver bream, which is lighter in colour.

The common bream is found in large shoals and is able to survive when oxygen levels are low in the summer months. It moves to deeper water in the winter. The spawning season lasts from May to July when the male, after developing tubercles on his body and fins, defends a territory that measures about five metres. The female then pairs with the male and lays her eggs at night over vegetation. Up to 340,000 eggs will be produced; they hatch after about two weeks. These fish mature at the age of five.

EEL (ANGUILLA ANGUILLA)

Eels are born in the Saragossa Sea, which is between Bermuda and the West Indies. The larvae are then swept by the Atlantic current and the Gulf Stream for three years before they metamorphose into elvers. British waterways are entered in January, if in the south-west, or April if in the north-east. The eel then spends between five and ten years in fresh water. It sometimes travels over land to move to different sections of water and is able to do so by closing its gills and keeping a supply of water in the large gill cavity. Its body eventually undergoes a series of changes to adapt to ocean life. The eyes grow in size, the intestine shrinks and the body becomes harder, with fat reserves. It then returns to the sea to spawn, after which it dies. Millions of eggs are left behind.

SIZE Length is between 42–100cm (female) and 29–51cm (male).

DESCRIPTION Snake-like appearance with no scales. Small pelvic fins near to the head. When they arrive in river systems they are brownish with yellow undersides; when they leave they are a grey colour with silver undersides.

HABITAT Any fresh water including lakes, ponds and estuaries.

FEEDING Invertebrates, small fish, frogs.

PIKE (ESOX LUCIUS)

The pike is a vicious predator that hunts alone. It lurks motionless amongst vegetation, waiting until its prey is close enough to reach. It breeds in April or May and is mature by the age of two or three. At breeding time, the female is courted by two or more males and will eventually lay up to 500,000 eggs which are shed simultaneously with the milt and stick to weeds after fertilisation. They hatch after two to three weeks and the resulting larvae also stick to the weeds using the large yolk sacs for nourishment. After five days they are able to swim away, feeding independently on plankton and small insects. Once they have reached three centimetres in length they begin to feed on small fish. From then on they are fully carnivorous, even eating other pike or fish the same size.

SIZE Length up to 130cm (female) 95cm (male).

DESCRIPTION Large fish with a long snout. Marbled brown and green markings with white undersides. Very large mouth with backward-pointing teeth and a long tail.

HABITAT Deep, calm water with an abundance of weeds such as lakes and slow-flowing rivers.

FEEDING Fish, small mammals and young water birds.

PERCH
(PERCA FLUVIATILIS)

The perch is a very popular fish with anglers. It inhabits shallow water in the spring and summer, moving to deeper water in the winter. The spawning season lasts from late April to June when the eggs are laid amongst plants and twigs. The female produces the eggs in long, sticky strands and they are covered by a thick mucus. After fertilisation this swells, protecting them. As many as 300,000 are laid, which hatch after seven to ten days. The young gather in shoals feeding on plankton.

SIZE Length usually 20–30cm but can reach up to 60cm.

DESCRIPTION Dark green upper sides with lateral bars across back. Two prominent dorsal fins with red pectoral, pelvic and anal fins.

HABITAT Likes lakes, ponds and slow-flowing rivers with clean water.

FEEDING Invertebrates and fish.

SIMILAR SPECIES Ruffe, which is smaller with black spots on its back and sides. The perch often hybridises with the ruffe.

TROUT (SALMO TRUTTA)

All trout spawn in the autumn and winter, with sea trout travelling up from the sea and lake trout moving towards feeder streams. The female makes a nest in the gravel up to ten centimetres deep where the male then joins her. Sperm and eggs are released; fertilization is simultaneous. As many as several hundred eggs are laid, which the female covers with gravel. The fish then move back downstream or return to the sea. The sea trout, unlike the Atlantic salmon, continues to feed while in the river, so there is not the same high mortality after spawning. Eggs develop at different rates, depending on the temperature of the water, but generally those laid in mid-November will hatch in late February or early March. The young are called alevins and hide in the gravel for up to six weeks using their yolk sacs for nourishment. They then come to the surface to feed and establish their own feeding territory of about twenty square centimetres, at which stage they are called trout fry. They eat insect larvae at first but invertebrates such as beetles and earthworms become more important in their second year. As they increase in size they move on to snails and shrimps and, finally, other fish.

SIZE Length between 25–40cm but can reach up to 60cm.
DESCRIPTION Blunt head with eye positioned at rear of mouth. Dotted markings on upper body. Undersides are silvery but colours of upper body will vary as there are many forms or sub-species that are determined by the environment.
HABITAT Sea trout prefers cold, well-oxygenated, fast-flowing streams for spawning but will also pass slower rivers on migration. Brown trout prefers clean freshwater lakes and rivers.
SIMILAR SPECIES Salmon, which is much larger with slimmer wrist at tail.

INDEX

ACKNOWLEDGEMENTS

Illustrations

The publisher would like to thank Oxford Scientific/Photolibrary.com for providing the photographs for this book. We would also like to thank the following for their kind permission to reproduce their photographs:

15, 54/55 Marianne Wilding; 16, 28/29, 36/37 (and front cover image), 42, 43, 48/49, 58/59, 66, 68/69, 80/81, 84/85, 86, 94, 98/99, 104, 107, 109, 116 (and back flap image), 122, 126, 128/129, 134, 135, 136/137, 138, 139, 146, 147, 148/149, 150, 151, 162, 166/167 Mark Hamblin; 17, 57, 92, 119 Chris Knights; 18/19, 20, 22/23, 78, 110/111, 112, Michael Leach; 21, 130 Robin Redfern; 25 P Kent; 27, 63, 83, 93, 118, 124/125, 131, 132 Tony Tilford; 32/33, 71 Ian West; 34 Liz & Tony Bomford; 38/39 Richard Day; 46 Roland Mayer; 40/41,108 Norbert Rosing; 47 Daniel G Cox; 30/31 50/51, 141 Niall Benvie; 52/53 (and back cover image) , 212 David Cayless; 56, 143, 238, 244/245, 247, 248/249 Hans Reinhard; 60/61 Stepen Dalton; 67, 100, 123 Richard Packwood; 73 Tom Ulrich; 74/75, 176, 178/179, 211, 215, 225, 181, 196 Harold Taylor Abipp; 79, 95 Eric Woods; 82 Michael Sewell; 87, 117 Alan Hartley; 88/89, 140 D J Sanders; 70, 90, 96, 101, 105, 114/115 David Tipling; 91 Jo McDonald/Okapia; 120, 142, 240/241 Keith Ringland;127 Jos Korenromp; 133 Terry Andrewartha; 144/145 Ben Osborne; 156/157, 158/159, 164/165 Paul Franklin; 174/175 David M Morris; 173 A L Cooke; 177 Jo Frohlich/Okapia; 180 R L Manuel; 182, 190/191, 206 Larry Crowhurst; 183 John Woolmer 184, 197 G McClean; 185, 230/231 Alastair Shay; 188 Frithjof Skibbe ; 189 JS & EJ Woomer; 192/193 Dennis Green; 194, 202,220, 250/251 OSF; 199, 205 David Fox; 204 Mike Linley; 207 Harry Fox; 208/209 Owen Newman; 210 Bob Fredrick; 212 Archie Allnutt; 213 Arthur Butler; 216/217 David Thompson; 218 K G Volk/Okapia; 219, Mike Birkhead; 221 Derek Bromhall; 222 Paulo De Oliveria; 62, 223, 228 David Boag; 224, 229, 232, 234, 243 Colin Milkins; 233 Martyn Chilmaid; 235 M Dennis; 239, 252/253 Max Gibbs; 242 Peter Gathercote; 76/77, 176 Barry Walker; 214 D G Fox; 226/227 G I Bernard; 97 Kenneth Day; 163, 160 Manfred Pfefferle; 246 Merlet; 161 Paul Franklin; 72 Edward Robinson; 200/201 Patti Murray;35 Barrie Watt;102/103 Mike Powles; 170 Konrad Wothe; 44/45 Berndt Fischer

With thanks to
Oliver Higgs, Kate Santon, Kate Truman,
Ruth Blair and everyone at Oxford Scientific Films